Andreas Fritsch

Elasticity and strength of bone and bone replacement materials

Andreas Fritsch

Elasticity and strength of bone and bone replacement materials

A multiscale continuum micromechanics approach

Südwestdeutscher Verlag für Hochschulschriften

Impressum/Imprint (nur für Deutschland/ only for Germany)
Bibliografische Information der Deutschen Nationalbibliothek: Die Deutsche Nationalbibliothek verzeichnet diese Publikation in der Deutschen Nationalbibliografie; detaillierte bibliografische Daten sind im Internet über http://dnb.d-nb.de abrufbar.

Verlag: Südwestdeutscher Verlag für Hochschulschriften Aktiengesellschaft & Co. KG
Dudweiler Landstr. 99, 66123 Saarbrücken, Deutschland
Telefon +49 681 37 20 271-1, Telefax +49 681 37 20 271-0, Email: info@svh-verlag.de
Zugl.: Wien, TU, Diss., 2009

Herstellung in Deutschland:
Schaltungsdienst Lange o.H.G., Berlin
Books on Demand GmbH, Norderstedt
Reha GmbH, Saarbrücken
Amazon Distribution GmbH, Leipzig
ISBN: 978-3-8381-0507-9

Imprint (only for USA, GB)
Bibliographic information published by the Deutsche Nationalbibliothek: The Deutsche Nationalbibliothek lists this publication in the Deutsche Nationalbibliografie; detailed bibliographic data are available in the Internet at http://dnb.d-nb.de.

Publisher:
Südwestdeutscher Verlag für Hochschulschriften Aktiengesellschaft & Co. KG
Dudweiler Landstr. 99, 66123 Saarbrücken, Germany
Phone +49 681 37 20 271-1, Fax +49 681 37 20 271-0, Email: info@svh-verlag.de

Printed in the U.S.A.
Printed in the U.K. by (see last page)
ISBN: 978-3-8381-0507-9

Contents

Abstract

Bone is a hierarchically organized material, characterized by an astonishing variability and diversity. Bone replacement or biomaterials are critical components in artificial organs, and they are also used as scaffolds in tissue engineering. The aim of this thesis is the prediction of the strength of bone and bone replacement materials, from their composition and microstructure, by means of multiscale models. The theoretical developments are supported by comprehensive experiments on cortical bone and on biomaterials made of hydroxyapatite, glass-ceramic, and titanium.

Chapter A investigates different morphological concepts (spheres vs. needles) for homogenization of linear elastic properties of porous polycrystals, as can be found in the mineral phase of bone.

Chapter B proposes a first attempt to model the strength properties of hydroxyapatite biomaterials, based on a micromechanical description of the elasticity and brittle failure of interfaces between isotropic, spherical crystals. In order to avoid optimization procedures for back-analysis of interface properties (as used in Chapter B), we developed an alternative approach **(Chapter C)** where we considered the non-spherical shape of the hydroxyapatite crystals. Using needles implies a 1D stress state in the bulk phase related to the needle direction, and this stress can be regarded as relevant for the stresses at the interface between crystals.

Chapter D presents an experimentally supported micromechanical explanation of cortical bone strength, based on a new vision on bone material failure: mutual ductile sliding of hydroxyapatite mineral crystals along layered water films is followed by rupture of collagen crosslinks. The multiscale micromechanics model is shown to be able to satisfactorily predict the strength characteristics of different bones from different species, on the basis of their mineral/collagen content, their porosities, and the elastic and strength properties of hydroxyapatite and (molecular) collagen.

Experimental investigations and modeling of two other classes of biomaterials accompany the theoretical developments: In **Chapter E**, porous titanium samples are tested acoustically and mechanically, and the corresponding mechanical properties, stiffness and strength, are predicted by a poro-micromechanical model. **Chapter F** presents a micromechanical description of bioresorbable porous glass ceramic scaffolds. Again, a material model predicting relationships between porosity and elastic/strength properties is developed and validated.

Introductory remarks

Presentation of investigated materials

Bone

Bone materials are characterized by an astonishing variability and diversity. Their hierarchical organizations are often well suited and seemingly optimized to fulfill specific mechanical functions. This has motivated research in the fields of bionics and biomimetics. The aforementioned optimization is primarily driven by selection during the biological evolution process. However, apart from the fact that selection is quite unlikely to push bone skeletal and material design to a well-defined optimum (Nowlan and Prendergast 2005), it is of great importance to notice that selection is realized at the level of the individual plant or animal (and not at the material level). Therefore, material optimization in the strictest sense of the word does not take place. Rather, 'architectural constraints' (Seilacher 1970; Gould and Lewontin 1979) merely due to once chosen material constituents and their physical interactions imply the fundamental hierarchical organization patterns or basic building plans, which remain largely unchanged during biological evolution. These building plans are expressed by typical morphological features which can be discerned across all bone materials. Katz et al. (1984) distinguish five levels of hierarchical organization, which have been quite generally accepted in the scientific community:

- The macrostructure at an observation scale of several mm to cm, where cortical (or compact) bone and trabecular (or spongy) bone can be distinguished [Fig. 1(a) and (b)];
- The microstructure at an observation scale of several 100 μm to several mm, where cylindrical units called osteons build up cortical bone, and where the single trabecular struts or plates can be distinguished [Fig.1(c) and (d)];
- The ultrastructure (or extracellular solid bone matrix) at an observation scale of several μm, comprising the material building up both trabecular struts and osteons [Fig.1(e)].
- Within the ultrastructure, collagen-rich domains [light areas in Fig.1(e)] and collagen-free domains [dark areas in Fig.1(e)] can be distinguished at an observation scale of several hundred nanometers. Commonly, these domains are referred to as fibrils and extrafibrillar space.

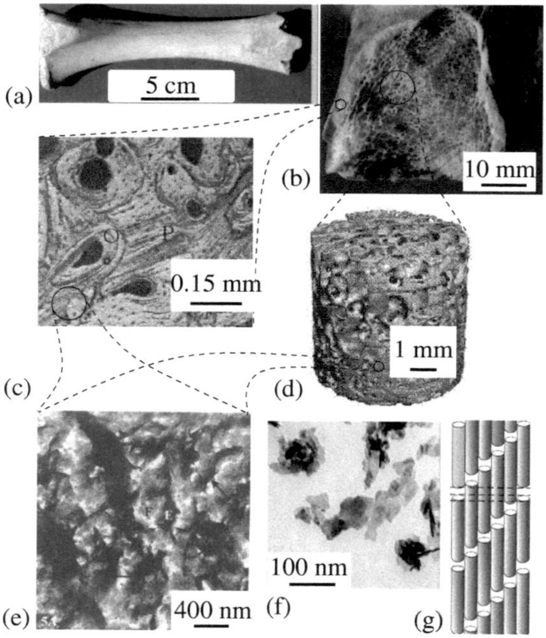

Figure 1: Hierarchical organization of bone: (a) whole long bone (macrostructure)(+); (b) section through long bone (macrostructure)(+); (c) osteonal cortical bone (microstructure)(o); (d) trabecular spaceframe (microstructure)(□); (e) ultrastructure(×); (f) hydroxyapatite crystals (elementary components)(+); (g) collagen molecules (elementary components)(+); (+)... From (Weiner and Wagner 1998), reprinted, with permission, from the Annual Review of Materials Science 28, ©1998 by Annual Reviews, www.annualreviews.org; (o)... Reprinted with permission from Lees et al. (1979a). ©1979, American Institute of Physics; (□)... reprinted from (Ding and Hvid 2000), with permission from Elsevier; (×) ... With kind permission from Springer Science+Business Media: (Prostak and Lees 1996, p.478, Fig. 5a).

- Finally, at an observation scale of several ten nanometers, the so-called elementary components of mineralized tissues can be distinguished. These are
  - Plate or needle-shaped mineral crystals consisting of impure hydroxyapatite (HA; $Ca_{10}[PO_4]_6[OH]_2$) with typical 1 to 5 nm thickness, and 25 to 50 nm length (Weiner and Wagner 1998) [Fig.1(f)];
  - Long cylindrically shaped collagen molecules with a diameter of about 1.2 nm and a length of about 300 nm (Lees 1987a), which are self-assembled in staggered organizational schemes (fibrils) with characteristic diameters of 50 to 500 nm (Cusack and Miller 1979; Miller 1984; Lees et al. 1990, 1994; Weiner et al. 1997; Weiner and Wagner 1998; Rho et al. 1998; Prostak and Lees 1996), [Fig.1(g)]; several covalently bonded fibrils are sometimes referred to as fibers;
  - Different non-collagenous organic molecules, predominantly lipids and proteins (Urist et al. 1983; Hunter et al. 1996); and
  - Water.

The present thesis extends a previously published multi-scale model for bone elasticity (Fritsch and Hellmich 2007) to bone strength, with emphasis on the material 'cortical bone' (see Chapter D).

Biomaterials and tissue engineering scaffolds

Biomaterials are critical components in artificial organs, and they are also used as scaffolds in tissue engineering (see next paragraph for more details). Biomaterial production includes metals, ceramics, polymers, and biocomposites. Metals such as stainless steel, cobalt alloys, titanium and titanium alloys are preferred for orthopedic applications due to their high strength and toughness. Ceramics are solid materials composed of inorganic, non-metallic substances. They are produced at high temperatures above 500°C and are characterized by their brittleness and high hardness. Bioceramics are used for implants and in the repair and reconstruction of diseased or damaged body parts. Examples of bioceramics are alumina, zirconia, titania, tricalcium phosphate, hydroxyapatite, calcium aluminates, bioactive glasses and glass-ceramics.

Tissue engineering is the laboratory-based design and construction of living, functional components that can be used for the regeneration of malfunctioning tissues (Buttery and Bishop 2005). Ideally, stem cells are extracted from a patient, seeded on a scaffold *in vitro*, and with the help of biological signals a tissue will grow. In more detail, the term scaffold refers to a structure, realized with natural or synthesized materials, which is able to promote cellular regeneration and to guide bone regeneration. Therefore, synthetic scaffolds may be seeded with carefully chosen biological cells and/or growth factors. Within this concept, the main role of a scaffold is to assure a mechanical support to the growing tissue, to guide this growth and to induce correct development of the bony organ. Due to their stimulating effects on bone cells, ceramics (such as hydroxyapatite, β-tricalcium phosphate, bioactive glasses, or glass ceramics) are identified as expressly promising materials for fabrication of tissue engineering scaffolds.

However, the design of such scaffolds is still a great challenge since (at least) two competing requirements must be fulfilled:

1. on the one hand, the scaffold must exhibit a sufficient mechanical competence, i.e. stiffness and strength comparable to natural bones;

2. on the other hand, once the scaffold would be implanted into the living organism, it should be continuously resorbed and replaced by natural bones. This typically requires a sufficient pore space (pore size in the range of hundred micrometers and porosity of more than 50-60% (Cancedda et al. 2007)), which discriminates the aforementioned mechanical properties, and therefore competes with the first requirement.

As concerns biomaterials, the present work focuses on modeling the macroscopic mechanical properties (elasticity and strength) of hydroxyapatite biomaterials as their properties are very similar to those of one major component of natural bone, namely bone mineral (see Chapters B and C). In particular the third paper (Chapter C) lays the foundation for a micromechanical description of the extracellular mineral, relevant for bone (dealt with in Chapter D).

In addition, mechanical characterization through acoustic, uniaxial, and triaxial testing as well as application of micromechanical models is shown for porous titanium biomaterials (see Chapter E) and porous glass-ceramic scaffolds (see Chapter F).

Hypotheses and limits

Morphology

The real morphology of bone mineral crystals is still an open question. Observations with atomic force microscopy (Eppell et al. 2001; Tong et al. 2003; Hassenkam et al. 2004), scanning electron microscopy (SEM) and transmission electron microscopy (TEM) (Traub et al. 1989; Su et al. 2003) reveal a rather plate-shaped morphology, being in contrast to a needle-like crystal shape observed with TEM (Lees et al. 1994) or X-ray small angle scattering (Fratzl et al. 1996).

The same ambiguity can be found for artificially produced hydroxyapatite biomaterials. There is evidence for spherical crystals from SEM (De With et al. 1981; Liu 1997), but also for rather elongated morphologies (Martin and Brown 1995). These hydroxyapatite ceramics are typically produced by sintering at temperatures above 500°C with resulting crystal size in the micrometer range. There are only few attempts to synthesize hydroxyapatite at physiological temperatures (Martin and Brown 1995; Tadic and Epple 2003), and only the latter study produced nanosized crystals.

Given the absence of a confirmed morphological description of hydroxyapatite crystals in artificial biomaterials as well as in natural bone, different hypotheses were tested. The aim was to identify a morphological description being sufficient for prediction of the mechanical properties of both materials.

In Chapter B, hydroxyapatite biomaterials are envisioned as porous polycrystals with a non-porous matrix. This matrix consists of *spherical* crystals with weak interfaces. A second approach is presented in Chapter C: Based on the morphological description of a polycrystal developed in Chapter A, hydroxyapatite biomaterials are represented as a polycrystal consisting of uniformly distributed crystal *needles* and spherical pores. The experimental validation for elasticity and strength indicates the superiority of the latter model.

Brittle versus ductile behavior of crystals

In Chapter C, a brittle behavior of the hydroxyapatite crystal needles within biomaterials is considered, whereas in Chapter D, we propose a (layered water-induced) ductile behavior for interfaces between the hydroxyapatite crystals as part of natural collagenous bone tissue. The reason for the different behaviors may well lie in the characteristic size of the crystals, and hence of the nature of their contact surfaces, the crystals in collagenous bone tissue being much smaller than the biomaterial crystals. In the same sense, in low or non-collagenous tissues, such as specific whale bones (Zioupos et al. 1997), the minerals grow larger, and also these tissues exhibit a brittle failure behavior. The idea of increased ductility due to increased activity of layered water films is also supported by the fact (Nyman et al. 2008) that bound water content is correlated to bone toughness; and this idea fits well with the suggestions of Boskey (2003), that larger crystals (implying less layered water films per crystal content) would lead to a more brittle behavior of bone materials.

Mechanical properties of elementary constituents

Validation of the micromechanical predictions for macroscopic mechanical properties (elasticity and strength) of bone and biomaterials is based on 'universal' micro/nanoscopic mechanical properties of the elementary constituents of the considered material. These properties are tissue and biomaterial-independent, and they are derived from experimental investigations. These 'universal' properties are the stiffness and strength characteristics of hydroxyapatite crystals and their interfaces (see Chapters B and C for the case of artificial biomaterials as well as Chapter D for the case of natural bone), of (molecular) collagen and of water (see Chapter D), of pure titanium (see Chapter E for the case of metallic biomaterials), and of a dense glass ceramic matrix (see Chapter F for ceramic biomaterials).

Concerning the tissue-independent elastic phase properties of bone (Chapter D), we consider the following experiments: Tests with an ultrasonic interferometer coupled with a solid media pressure apparatus (Katz and Ukraincik 1971; Gilmore and Katz 1982) reveal the isotropic elastic properties of hydroxyapatite powder, which, in view of the largely disordered arrangement of minerals (Lees et al. 1994; Fratzl et al. 1996; Peters et al. 2000; Hellmich and Ulm 2002a), are sufficient for the characterization of the mineral phase (Hellmich and Ulm 2002b; Hellmich et al. 2004b; Fritsch et al. 2006). Given the absence of direct measurements of (molecular) collagen, its elastic properties are approximated by those of dry rat tail tendon, a tissue consisting almost exclusively of collagen. By means of Brillouin light scattering, Cusack and Miller (1979) have determined the respective five independent elastic constants of a transversely isotropic material (Table D.1). We assign the standard bulk modulus of water (Table D.1) to phases comprising water with mechanically insignificant non-collageneous organic matter.

Concerning the biomaterial-independent elastic properties of artificial hydroxyapatite crystal (Chapters B and C) we adapt those chosen for bone mineral.

The approach proposed in Chapter B relies on three 'universal' material properties of interfaces between single hydroxyapatite crystals represented, for mathematical tractability, as spheres. The interface properties are difficult to be directly accessed, namely the friction angle α, the cohesion h, and a dimensionless quantity κ of the interfaces. Therefore, these phase properties are determined by means of an optimization procedure providing the closest match of model predictions to experimentally determined uniaxial compressive strength data of hydroxyapatite biomaterials. Applying an evolution algorithm yields a set of solution vectors which are equal in terms of the highly satisfactory correlation coefficient between the respective model predictions and the corresponding experimental data for uniaxial compressive strength (see Section B.5.4).

In order to avoid such an optimization procedure for back-analysis of interface properties, we developed an alternative approach where we considered the non-spherical shape of the hydroxyapatite crystals. Using needles suggests a predominant stress state related to the needle direction, and given this virtual 1D situation, this stress can be regarded as relevant for the stresses at the interface between crystals. In this sense, the approach proposed in Chapter C relies on the strength properties of interfaces between needle-shaped hydroxyapatite crystals, expressed by the bulk phase 'hydroxyapatite', namely its tensile and shear strength, $\sigma_{HA}^{ult,t}$ and $\sigma_{HA}^{ult,s}$. We are not aware of direct strength tests on pure hydroxyapatite (with $\phi = 0$). Therefore, we consider one uniaxial tensile test and one uniaxial compressive test on the densest samples available. From these two tests, we back-calculate the universal tensile and shear strength of pure hydroxyapatite relevant for crystal interfaces (Table C.2). It is interesting to note that consideration of the normal stress alone proved to be not sufficient for predicting macroscopic failure, in particular for low porosities in uniaxial compression. Only the 'mixed' formulation of the failure criterion taking into account normal and shear stresses (see Section C.3.2) inside the needles delivers satisfying macroscopic strength predictions.

Experimental data for model validation

The micromechanical models presented in Chapters B-F are based on experimentally determined elasticity and strength properties of the elementary material components. The models predict, for each set of tissue or biomaterial-specific volume fractions (e.g. porosities), the corresponding tissue or biomaterial-specific elasticity and strength properties at all observation scales. Thus, a strict experimental validation of the mathematical model is realized as follows: (i) different sets of volume fractions are determined from composition experiments on different bone or biomaterial samples; (ii) these volume fractions are used as model input, and (iii) corresponding model-predicted stiffness and strength values (model output) are compared to results from stiffness and strength experiments on the same or very similar bone or biomaterial samples.

Elastic macroscopic properties of biomaterials can be determined through uniaxial quasi-static mechanical tests, ultrasonic techniques or resonance frequency tests (Chapters C, E and F). Typical sample geometries include cylinders (diameter 5 mm, length 10 mm) for titanium samples (Chapter E) and glass ceramic scaffolds (Chapter F), and millimeter or centimeter-sized cylinders, bars or discs for hydroxyapatite biomaterials (see Chapter C and Table C.1). In case of ultrasonic testing, the length of the propagating wave has to be taken into account: If the wavelength is considerably smaller than the diameter of the specimen, a (compressional) 'bulk wave', i.e. a laterally constrained wave, propagates in a quasi-infinite medium. On the other hand, if the wavelength is considerably larger than the diameter of the specimen, a 'bar wave' propagates, i.e. the specimen acts as one-dimensional bar without lateral constraints.

Macroscopic uniaxial strength properties of bone and biomaterial samples can be determined through quasi-static tensile, compressive and bending tests (Chapters C-F). Typical sample geometries include cylinders (diameter 5 mm, length 10 mm) for titanium samples (Chapter E) and glass ceramic scaffolds (Chapter F), millimeter or centimeter-sized cylinders, bars or discs for hydroxyapatite biomaterials (see Chapter C and Table C.1) and millimeter or centimeter-sized cylinders or parallelepipeds, often with reduced cross section, for bone (see Chapter D and Table D.4).

Original contributions to the field of micromechanical modeling

Effect of morphology in self-consistent schemes

The classical self-consistent scheme (Hershey 1954; Kröner 1958; Hill 1963) is often used for modeling the overall elastic properties of porous polycrystals. It consists in embedding spherical inclusions into a matrix with stiffness of the homogenized material. This approach predicts a vanishing overall stiffness ('percolation threshold') for porosities greater than 50%.

In Chapter A, it is proposed to replace spherical solid inclusions by a set of infinitely many uniformly oriented cylindrical inclusions (needles). All these needles are identical with respect to shape and material behavior, while being oriented in all directions in space. This has two implications: (i) the stiffness tensor related to a single crystal is a function of the Euler angles, while the components are orientation-independent in a local base frame, and (ii) the (overall) effective stiffness tensor of the porous polycrystal is isotropic.

Interfaces

Interfaces are often believed to play a role in the mechanical behavior of mineralized biological and biomimetic materials (Bhowmik et al. 2007). In Chapter B, porous hydroxyapatite biomaterials are represented as a (dense) polycrystal with weak interfaces, which serves as the skeleton of a porous material defined one observation scale above. In detail, isotropic single crystals of typically quasi-spherical shape are separated from each other by very thin (essentially 2D) interfaces. The interface stiffness tensor exhibits an infinite normal component and a positive tangential component, and its load bearing capacity is characterized by a Coulomb-type law, considering the tangential and normal components of the traction force acting on the interface (see Section B.3 for details). In order to determine the effective failure properties resulting from local (brittle) failure characteristics and from the interactions between interfaces and bulk single crystals, the local interface forces have to be related to the 'macroscopic' stresses. The tangential and normal traction forces occurring in the interface failure criterion are non-homogeneously distributed across the interfaces. Failure will occur where relatively high tangential traction forces encounter a relative low resistance due to relatively low normal traction forces. Instead of trying to model the actual force fields across the interfaces, we estimate the effect of the actual force distribution through so-called *effective* traction forces, as it is commonly done for stress, strain, or force fields in the context of continuum micromechanics (Suquet 1997a; Dormieux et al. 2007). In this line, we represent the failure-inducing interplay between moderate normal traction forces and tangential traction force *peaks* by means of two different *effective* measures for the normal and the tangential traction forces, respectively: (i) first-order moments of normal forces, and (ii) second-order moments (also called quadratic average) of tangential forces, in the line of (Kreher 1990; Kreher and Molinari 1993; Dormieux et al. 2002). The relation between the quadratic average and the macroscopic stress is established through energy considerations. Remarkably, the second-order moment of tangential tractions over all interfaces within the RVE is proportional to the 'macroscopic' equivalent deviatoric stress, and local, Coulomb-type brittle failure in the interfaces implies Drucker-Prager-type (brittle, elastic limit-type) failure properties at the scale of the polycrystal.

It is also interesting to note that the elastic, brittle failure criterion is quasi-identical to the yield surface of a porous medium obtained through non-linear homogenization (Dormieux 2005; Dormieux et al. 2006b) which is related to failure of a ductile solid matrix obeying a Drucker-

Prager criterion. The ductile criterion is even identical to the elastic domain for incompressible solid matrices, see Section B.5 for a detailed discussion.

Organization of the thesis

The overall aim of this thesis is the prediction of bone strength from its composition and microstructure. Classically, the strength of bone materials is thought to be related to the strength properties of hydroxyapatite and collagen, and/or interfaces between these constituents. Chapters B and C concentrate on the failure properties of artificial hydroxyapatite biomaterials which are very similar to natural bone mineral, based on a morphological concept presented in Chapter A. A micromechanical model for bone strength is presented in Chapter D, while some experimental investigations and modeling of biomaterials accompany the theoretical developments (Chapters E and F).

Chapter A is dedicated to the homogenization of linear elastic properties of porous polycrystals built up of needle-like platelets or sheets. Such microstructures can be found in a number of biological and man-made materials such as the mineral phase of bone, the cement paste of concrete or gypsum. Within a self-consistent scheme the solid phase is represented by cylindrical inclusions (needles). Uniform and axisymmetrical orientation distribution of linear elastic, isotropic as well as anisotropic needles is considered and the results are compared to the classical ones related to spherical inclusions. As a key result, a porosity lower than 0.4 is shown to result in the (overall) elastic properties of the polycrystal with uniformly oriented needles, which are quasi-identical to those of a polycrystal with solid spheres. However, as opposed to the sphere-based model, the needle-based model does not predict a percolation threshold for inclusions with infinite aspect ratio.

Chapter B proposes a first attempt to model the strength properties of hydroxyapatite biomaterials, based on a micromechanical description of the elasticity and brittle failure of interfaces between isotropic crystals in a (dense) polycrystal, which serves as the skeleton of a porous material defined one observation scale above. Equilibrium and compatibility conditions, together with a suitable matrix-inclusion problem with a compliant interface, yield the homogenized elastic properties of the polycrystal, and of the porous material with polycrystalline solid phase. Incompressibility of single crystals guarantees finite shear stiffness of the polycrystal, even for vanishing interface stiffness, while increasing the latter generally leads to an increase of polycrystal shear stiffness. Corresponding elastic energy expressions give access to effective stresses representing the stress heterogeneities in the microstructures, which induce brittle failure. Thereby, Coulomb-type brittle failure of the crystalline interfaces implies Drucker-Prager-type (brittle, elastic limit-type) failure properties at the scale of the polycrystal. At the even higher scale of the porous material, high interfacial rigidities or low interfacial friction angles may result in closed elastic domains, indicating material failure even under hydrostatic pressure. This micromechanics model can satisfactorily reproduce the compressive experimental strength data of different (brittle) hydroxyapatite biomaterials, across largely variable porosities. Thereby, the brittle failure criteria can be well approximated by micromechanically derived criteria referring to ductile solid matrices, both criteria being even identical if the solid matrix is incompressible.

A second approach for modeling the strength properties of hydroxyapatite biomaterials is addressed in **Chapter C**, with the aim to predict uniaxial compressive *and* tensile failure. Thereby, these biomaterials are envisioned as porous polycrystals consisting of (isotropic) hydroxyapatite needles and spherical pores, in the line of Chapter A. Failure possibly occurs at the interfaces of the crystal needles, but modeling interfaces between non-spherical objects is extremely complex. Therefore, the effect of 'micro'-interface behavior of elongated 1D particles on the overall 'macroscopic' material is mimicked by equivalent 'bulk' failure properties of the crystal needles. Validation of respective micromechanical models relies on two independent experimental sets: Biomaterial-specific macroscopic (homogenized) stiffness and uniaxial (tensile and compressive) strength predicted from biomaterial-specific porosities, on the basis of biomaterial-independent ('universal') elastic and strength properties of hydroxyapatite, are compared to corresponding biomaterial-specific experimentally determined (acoustic and mechanical) stiffness and strength values. The good agreement between model predictions and the corresponding experiments underlines the relevance of this approach.

Chapter D proposes an experimentally supported micromechanical explanation of cortical bone strength, based on a new vision on bone material failure: mutual ductile sliding of hydroxyapatite mineral crystals along layered water films is followed by rupture of collagen crosslinks. In order to cast this vision into a mathematical form, a multiscale continuum micromechanics theory for upscaling of elastoplastic properties is developed, based on the concept of concentration and influence tensors for eigenstressed microheterogeneous materials. The model reflects bone's hierarchical organization, in terms of representative volume elements for cortical bone, for extravascular and extracellular bone material, for mineralized fibrils and the extrafibrillar space, and for wet collagen. In order to get access to the stress states at the interfaces between crystals, the extrafibrillar mineral is resolved into an infinite amount of cylindrical material phases oriented in all directions in space in the line of Chapter C. The multiscale micromechanics model is shown to be able to satisfactorily predict the strength characteristics of different bones from different species, on the basis of their mineral/collagen content, their intercrystalline, intermolecular, lacunar, and vascular porosities, and the elastic and strength properties of hydroxyapatite and (molecular) collagen.

In **Chapter E**, titanium with different porosities, produced on the basis of metal powder and space holder components, is investigated as bone replacement material. For the determination of mechanical properties, i.e. strength of dense and porous titanium samples, two kinds of experiments were performed – uniaxial and triaxial tests. The triaxial tests were of poromechanical nature, i.e. oil was employed to induce the same pressure both at the lateral surfaces of the cylindrical samples and inside the pores. The stiffness properties were revealed by acoustic (ultrasonic) tests. Different frequencies give access to different stiffness components (stiffness tensor components related to high-frequency-induced bulk waves versus Young's moduli related to low-frequency-induced bar waves), at different observation scales; namely, the observation scale the dense titanium with around 100 μm characteristic length (characterized through the high frequencies) versus that of the porous material with a few millimeters of characteristic length (characterized through the low frequencies). Finally, the experimental results were used to develop and validate a poro-micromechanical model for porous titanium, which quantifies material stiffness and strength from its porosity and (in the case of the aforementioned triaxial tests) its pore pressurization state.

Chapter F presents a micromechanical description of bioresorbable porous glass ceramic scaffolds used for bone tissue engineering. Based on continuum micromechanics, a material model predicting relationships between porosity and elastic/strength properties is employed. The model, which mathematically expresses the mechanical behavior of a ceramic matrix (based on a glass system of the type SiO_2-P_2O_5-CaO-MgO-Na_2O-K_2O; called CEL2) in which interconnected pores are embedded, is carefully validated through a wealth of independent experimental data. The remarkably good agreement between porosity-based model predictions for the elastic and strength properties of CEL2-based porous scaffolds and corresponding experimentally determined mechanical properties underlines the great potential of micromechanical modeling for speeding up the biomaterial and tissue engineering scaffold development process – by delivering reasonable estimates for the material behavior, also beyond experimentally observed situations.

Publication A

Porous polycrystals built up by uniformly and axisymmetrically oriented needles: Homogenization of elastic properties (Fritsch et al. 2006)

Authored by Andreas Fritsch, Luc Dormieux, and Christian Hellmich
Published in *Comptes Rendus Mecanique*, Volume 334, pages 151–157

Porous polycrystal-type microstructures built up of needle-like platelets or sheets are characteristic for a number of biological and man-made materials. Herein, we consider (i) uniform, (ii) axisymmetrical orientation distribution of linear elastic, isotropic as well as anisotropic needles. Axisymmetrical needle orientation requires derivation of the Hill tensor for arbitrarily oriented ellipsoidal inclusions with one axis tending towards infinity, embedded in a transversely isotropic matrix; therefore, Laws' integral expression of the Hill tensor is evaluated employing the theory of rational functions. For a porosity lower 0.4, the elastic properties of the polycrystal with uniformly oriented needles are quasi-identical to those of a polycrystal with solid spheres. However, as opposed to the sphere-based model, the needle-based model does not predict a percolation threshold. As regards axisymmetrical orientation distribution of needles, two effects are remarkable: Firstly, the sharper the cone of orientations the higher the anisotropy of the polycrystal. Secondly, for a given cone, the anisotropy increases with the porosity. Estimates for the polycrystal stiffness are hardly influenced by the anisotropy of the bone mineral needles. Our results also confirm the very high degree of orientation randomness of crystals building up mineral foams in bone tissues.

A.1 Introduction

Porous polycrystal-type microstructures built up of needle-like platelets or sheets can be found in a number of biological and man-made materials; such as bone (Hellmich et al. 2004a; Hellmich and Ulm 2002a) or eggs (Silyn-Roberts and Sharp 1986), or at the cement paste level of concrete (Baroughel-Bouny 1994). We here deal with homogenization of their overall (linear) elastic properties, by means of self-consistent schemes. Thereby, the solid phase (needles) is represented by cylindrical inclusions (a cylinder being the limit case of a prolate spheroid with its long axis being very much larger than its spherical axis), and the (empty) pore inclusions (drained conditions) are spherical; extension to pressurized pores according to Chateau and Dormieux (2002) is straightforward. Subsequently, we consider (i) uniform, (ii) axisymmetrical orientation distribution of isotropic as well as anisotropic needles with elasticity tensor $\mathbb{C}_s$.

A.2 Uniform orientation distribution of needles

Uniformly oriented needles result in isotropic elastic properties of the polycrystal. The corresponding stiffness estimate $\mathbb{C}^{SCS}$ reads as

$$\begin{aligned}\mathbb{C}^{SCS} &= (1-\phi)\,\mathbb{C}_s :< [\mathbb{I} + \mathbb{P}^{SCS}_{cyl} : (\mathbb{C}_s - \mathbb{C}^{SCS})]^{-1} >: \\ &\quad \{(1-\phi) < [\mathbb{I} + \mathbb{P}^{SCS}_{cyl} : (\mathbb{C}_s - \mathbb{C}^{SCS})]^{-1} > +\phi\,(\mathbb{I} - \mathbb{P}^{SCS}_{sph} : \mathbb{C}^{SCS})^{-1}\}^{-1} \end{aligned} \quad \text{(A.1)}$$

with

$$< [\mathbb{I} + \mathbb{P}^{SCS}_{cyl} : (\mathbb{C}_s - \mathbb{C}^{SCS})]^{-1} >= \int_{\varphi=0}^{2\pi}\int_{\vartheta=0}^{\pi} [\mathbb{I} + \mathbb{P}^{SCS}_{cyl}(\vartheta,\varphi)\,(\mathbb{C}_s - \mathbb{C}^{SCS})]^{-1} \frac{\sin\vartheta\,\mathrm{d}\vartheta\,\mathrm{d}\varphi}{4\pi} \quad \text{(A.2)}$$

where $\mathbb{I}$, $I_{ijkl} = 1/2(\delta_{ik}\delta_{jl} + \delta_{il}\delta_{kj})$, is the fourth-order unity tensor, δ_{ij} is the Kronecker delta, ϕ denotes the porosity, $\mathbb{P}^{SCS}_{sph}$ and $\mathbb{P}^{SCS}_{cyl}$ are the fourth-order Hill tensors for spherical and cylindrical inclusions, respectively. The Hill tensor for spherical inclusions, $\mathbb{P}^{SCS}_{sph}$, is widely available in the open literature (Eshelby 1957; Suvorov and Dvorak 2002). The components of the Hill tensor for cylindrical inclusions embedded in an isotropic medium are given for a base frame coinciding with the long axis of the cylinder (Eshelby 1957). Transformation of Hill tensors related to differently oriented cylindrical inclusions, to one reference frame can be expressed by Euler angles ϑ and φ, rendering $\mathbb{P} = \mathbb{P}^{SCS}_{cyl}(\vartheta,\varphi)$ in Eqn.(A.2).

The numerical solution of (A.1) shows that the effective Young's modulus E^{SCS} is practically independent of the needles' Poisson's ratio ν_s.

The question arises whether uniform orientation of needles can be appropriately considered by representing the solid phase simply by spherical inclusions. The corresponding self-consistent estimate $\mathbb{C}^{SCS}$ for identical shape and orientation of inclusions reads as (see e.g. (Zaoui 1997a))

$$\mathbb{C}^{SCS} = (1-\phi)\,\mathbb{C}_s : \{\mathbb{I} + \mathbb{P}^{SCS}_{sph} : (\mathbb{C}_s - \mathbb{C}^{SCS})\}^{-1} \tag{A.3}$$

In case of an incompressible solid phase (with bulk modulus $k_s \rightarrow \infty$), (A.3) can be solved analytically:

$$\mu^{SCS} = \mu_s \frac{3(1-2\phi)}{3-\phi}, \quad k^{SCS} = \frac{4(1-\phi)}{3\phi}\mu^{SCS} \tag{A.4}$$

where k^{SCS} and μ^{SCS} are the effective bulk and shear moduli, and μ_s is the shear modulus of the isotropic solid. This scheme shows a percolation threshold exactly equal to $\phi = \frac{1}{2}$, for any value of the Poisson's ratio ν_s of the solid phase. As for a compressible solid phase, the homogenized Young's modulus E^{SCS} can still be approximated by the affine expression $E_s(1-2\phi)$ with an error of at most 4 % relative to the exact solution, i.e. E^{SCS} is quasi-independent of Poisson's ratio.

On the entire porosity range, $0 < \phi < 1$, the self-consistent stiffness estimates based on uniformly oriented solid needles are quasi-identical for both isotropic and anisotropic needle behavior [Fig. A.1 (a) and (b), see Fig. A.1(c) for elastic constants (Katz and Ukraincik 1971) of hydroxyapatite crystals building up porous foams in bone (Hellmich and Ulm 2002a)]. In addition, on the interval $0 < \phi < 0.4$, these estimates are quasi-identical to those based on isotropic solid spheres [Fig. A.1 (a) and (b)]. From a physical viewpoint, one may argue that, at a sufficiently high concentration, both spherical as well as isotropic or anisotropic needle-type particles build up similar contiguous matrices. Particularly, in the vicinity of $\phi = 0$, the first-order expansions of the homogenized elastic constants with respect to the porosity are identical for the two models with an isotropic solid phase, reading as:

$$\frac{E^{SCS}}{E_s} = 1 - \frac{3}{2}\frac{(1-\nu_s)(5\nu_s+9)}{7-5\nu_s}\phi \quad ; \quad \nu_{SCS} = \nu_s + \phi\frac{3(1-5\nu_s)(1-\nu_s^2)}{2(7-5\nu_s)} \tag{A.5}$$

$$\frac{k^{SCS}}{k_s} = 1 - \frac{3}{2}\frac{1-\nu_s}{1-2\nu_s}\phi \quad ; \quad \frac{\mu^{SCS}}{\mu_s} = 1 - 15\frac{1-\nu_s}{7-5\nu_s}\phi \tag{A.6}$$

However, as opposed to the sphere-based model, the needle-based model does not predict any percolation threshold, i.e. E^{SCS}, k^{SCS} and $\mu^{SCS} \rightarrow 0$ only if the volume fraction of the solid phase becomes very small ($\phi \rightarrow 1$). From an intuitive viewpoint, this is consistent with the 'rice grain effect': As compared to spheres, needles are more likely to contact each other, especially at low volume fraction ($\phi \rightarrow 1$). A first-order expansion in the vicinity of $\phi = 1$ of μ^{SCS} (resp. k^{SCS}) can be sought in the form $\mu^{SCS} \sim m(1-\phi)$ [resp. $k^{SCS} \sim k(1-\phi)$]. As regards isotropic needles, analytical expressions for m and k can be derived and proven to be independent of ν_s :

$$m = \frac{71 - 2\sqrt{79}}{1575}, \quad k = \frac{-8 + 2\sqrt{79}}{189} \tag{A.7}$$

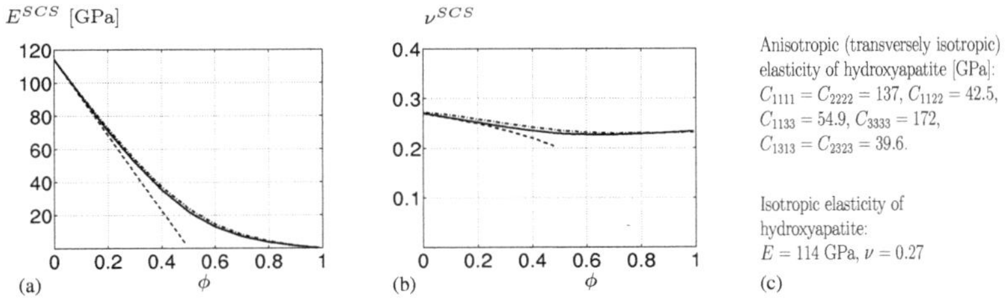

Figure A.1: (a) Young's modulus and (b) Poisson's ratio of isotropic porous polycrystals, predicted by the sphere-based and needle-based models, respectively (isotropic spheres ... dashed lines, uniformly oriented isotropic needles ... solid lines, uniformly oriented anisotropic needles ... dash–dot lines); (c) Anisotropic and isotropic elasticity of hydroxyapatite (Katz and Ukraincik 1971).

Accordingly, the limit of ν^{SCS} when ϕ tends towards 1 is independent of ν_s as well :

$$\lim_{\phi \to 1} \nu^{SCS} = \frac{17 - \sqrt{79}}{35} \tag{A.8}$$

A.3 Axisymmetric orientation distribution of needles

Axisymmetrically oriented needles result in transversely isotropic elastic properties of the polycrystal. With ϑ being measured with respect to the symmetry axis of the orientation distribution, we consider (i) uniform needle distribution in the cone $[0, \vartheta_{max}]$, and (ii) Gaussian needle distribution around $\vartheta_{max}/2$ with standard deviation s_ϑ; both expressed in terms of a distribution function $F(\vartheta)$. The corresponding stiffness estimate still obeys (A.1), while (A.2) now reads as

$$< (\mathbb{I} + \mathbb{P}^{SCS}_{cyl} : \delta\,\mathbb{C})^{-1} >= \int\limits_{\varphi=0}^{2\pi} \int\limits_{\vartheta=0}^{\vartheta_{max}} F(\vartheta)\,[\mathbb{I} + \mathbb{P}^{SCS}_{cyl}(\vartheta,\varphi)\,(\mathbb{C}_s - \mathbb{C}^{SCS})]^{-1} \frac{\sin\vartheta\,\mathrm{d}\vartheta\,\mathrm{d}\varphi}{2\pi(1-\cos\vartheta_{max})} \tag{A.9}$$

and while the Hill tensors $\mathbb{P}^{SCS}_{cyl}$ and $\mathbb{P}^{SCS}_{sph}$ now refer to inclusions in a transversely isotropic material.

Expressions for $\mathbb{P}^{SCS}_{sph}$ can be found in Hellmich et al. (2004a), and for determination of $\mathbb{P}^{SCS}_{cyl}$ we evaluate Laws' double integral expression of the Hill tensor (Laws 1977) for arbitrarily oriented cylindrical inclusions embedded in a transversely isotropic material, employing the theory of rational functions. Thereby, we arrive at a single-integrated expression allowing for efficient computational evaluation (see Appendix).

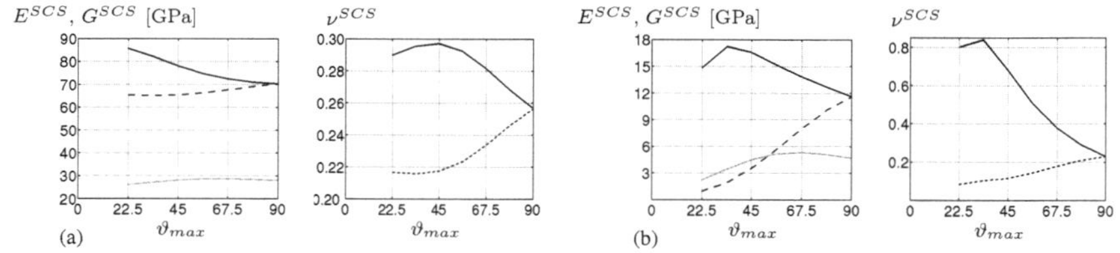

Figure A.2: Effect of axisymmetric distribution of anisotropic needles (uniformly distributed between $\vartheta = 0$ and $\vartheta = \vartheta_{max}$) on the longitudinal and transverse Young's moduli, Poisson's ratios, and shear modulus for different porosities [(a) $\phi = 0.2$, (b) $\phi = 0.6$]. Longitudinal components are shown as solid lines, transversal components as dashed lines, and the shear modulus as dotted line.

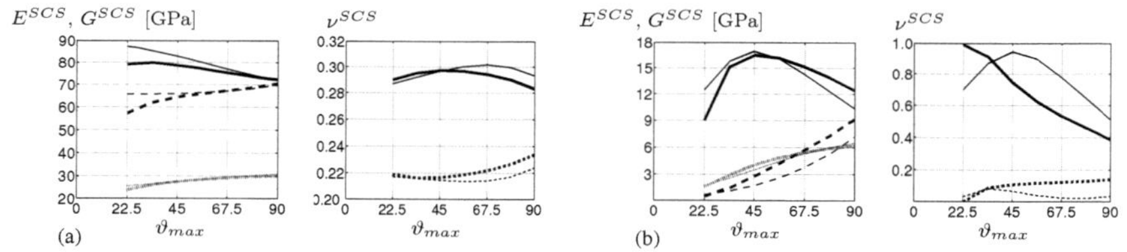

Figure A.3: Effect of axisymmetric distribution of anisotropic needles (Gaussian–type distributed around $\vartheta_{max}/2$ with standard deviation s_ϑ) on the longitudinal and transverse Young's moduli and Poisson's ratios for different porosities [(a) $\phi = 0.2$, (b) $\phi = 0.6$] and different standard deviations ($s_\vartheta = 2.5^o$ ...thick lines, $s_\vartheta = 12^o$ ...thin lines). Longitudinal components are shown as solid lines, transversal components as dashed lines, and the shear modulus as dotted line.

We evaluated Eqn.(A.9) for a uniform distribution of needles between 0 and a maximum angle ϑ_{max} as well as for a Gaussian distribution with different standard deviations around $\vartheta_{max}/2$, see Fig. A.2 and A.3. Two effects are remarkable (Fig. A.2): Firstly, as expected, the sharper the cone of orientations the higher is the anisotropy of the polycrystal. Secondly, the higher the porosity the more pronounced is the effect of the non-uniform needle orientation distribution, on both the Young's modulus and the Poisson's ratio. As compared to uniform needle distribution between $\vartheta = 0$ and $\vartheta = \vartheta_{max}$, the Gaussian distribution around $\vartheta_{max}/2$ with standard deviation s_ϑ significantly affects the effective Poisson's ratio (compare Fig. A.2 and A.3), while differences in Young's and shear moduli are, on the average, less than 7 % for the investigated distributions (Fig. A.2 and A.3).

A.4 Discussion

The present results are also noteworthy from a biomechanical viewpoint: In the ultrastructure of bones and mineralized tissues hydroxyapatite crystals build up a contiguous network or mineral foam (Hellmich and Ulm 2002a; Hellmich et al. 2004a). Single crystals have typical dimensions of 50 nm average length, 25 nm average width, and 1 to 7 nm thickness (Weiner and Wagner 1998; Fratzl et al. 1996). In a first approximation, they are often characterized as needles (Fratzl et al. 1996; Sasaki 1991; Fratzl et al. 1991). This renders the homogenization schemes developed here as appropriate for mineral foams occuring in bones. In particular, agreement between homogenized elastic properties of uniformly oriented needles with those of spheres for a porosity lower 0.4 (Fig. A.1) confirms the use of self-consistent schemes with spherical inclusions for hydroxyapatite polycrystals (Hellmich et al. 2004a), which have been validated by the experimental data of (Lees et al. 1983; Lees 1987a). At higher porosities, however, the needle-based scheme seems to be superior to the sphere-based scheme, since the former accounts for contiguity of the crystals, leading to non-zero homogenized stiffness, while the latter exhibits a percolation threshold beyond which the homogenized stiffness vanishes. Indeed, elasticity experiments (Lees and Page 1992) reveal that mineral crystals do contribute to the overall stiffness of low-mineralized turkey leg tendon, with a mineral foam porosity larger than 50%.

The present results also confirm the pronounced randomness of crystal orientation in bone tissues, revealed already by chemical (Peters et al. 2000) or mechanical (Hellmich and Ulm 2002a) means: Any pronounced orientation of needles leads to high anisotropy ratios E_{tran}/E_{long} far beyond two, and up to ten (Fig. A.2). In real bone ultrastructure, however, this ratio lies always markedly below two (Lees et al. 1979b, 1983; Hellmich and Ulm 2002a).

A.5 Appendix: Hill tensor for arbitrarily oriented cylindrical inclusions embedded in a transversely isotropic material

The starting point is Laws' classical expression for the Hill tensor (see for instance (Laws 1977, 1985)) :

$$\mathbb{P} = \frac{\omega_2\,\omega_3}{4\pi} \int\limits_{|\xi|=1} \frac{\mathbf{\Gamma}}{(\boldsymbol{\xi}\cdot\mathbf{A}^T\cdot\mathbf{A}\cdot\boldsymbol{\xi})^{3/2}}\,\mathrm{d}S(\boldsymbol{\xi}) \tag{A.10}$$

$\boldsymbol{\xi}$ is the unit length vector pointing from the origin of the sphere to the surface element $\mathrm{d}S(\boldsymbol{\xi})$. The second-order tensor $\mathbf{A}$ describes the shape of the ellipsoid, with base vectors $\mathbf{w}_1, \mathbf{w}_2$ and $\mathbf{w}_3$ pointing in the principal directions of the ellipsoid,

$$\mathbf{A} = \mathbf{w}_1 \otimes \mathbf{w}_1 + \omega_2\, \mathbf{w}_2 \otimes \mathbf{w}_2 + \omega_3\, \mathbf{w}_3 \otimes \mathbf{w}_3, \quad \omega_3 \gg 1 \tag{A.11}$$

The fourth-order tensor $\mathbf{\Gamma}$ is defined as

$$\mathbf{\Gamma} = \boldsymbol{\xi} \overset{s}{\otimes} \mathbf{K}^{-1} \overset{s}{\otimes} \boldsymbol{\xi}, \quad \mathbf{K} = \boldsymbol{\xi} \cdot \mathbb{C} \cdot \boldsymbol{\xi} \tag{A.12}$$

The second-order tensor $\mathbf{K}$ is the acoustic tensor, $\mathbb{C}$ is the stiffness tensor of the transversely isotropic matrix. $\overset{s}{\otimes}$ denotes the symmetrized tensor product.

The technique presented hereafter adapts the ideas presented in (Gruescu et al. 2005) and (Suvorov and Dvorak 2002) to cylindrical inclusions. First, we consider the denominator of expression (A.10). The unit vector $\boldsymbol{\xi}$ can be expressed in spherical coordinates $\Phi \in [0, 2\pi]$ and $\Theta \in [0, \pi]$ as $\xi_1 = \sin\Theta\, \cos\Phi$, $\xi_2 = \sin\Theta\, \sin\Phi$ and $\xi_3 = \cos\Theta$, so that $\mathrm{d}S = \sin\Theta\, \mathrm{d}\Phi\, \mathrm{d}\Theta$. Since

$$\boldsymbol{\xi} \cdot \mathbf{A}^T \cdot \mathbf{A} \cdot \boldsymbol{\xi} = \omega_3^2 \cos^2\Theta + \sin^2\Theta\, (\cos^2\Phi + \omega_2^2 \sin^2\Phi)$$

we find with $x = \cos\Theta$ and $\gamma^2 = \frac{1}{\omega_3^2}(\cos^2\Phi + \omega_2^2 \sin^2\Phi)$

$$\mathbb{P} = \frac{\omega_2}{4\pi} \int_0^{2\pi} \int_{-1}^{1} \frac{\gamma^2}{[x^2 + (1 - x^2)\gamma^2]^{3/2}}\, \frac{\mathbf{\Gamma}(x, \Phi)}{\cos^2\Phi + \omega_2^2 \sin^2\Phi}\, (-\mathrm{d}x)\, \mathrm{d}\Phi \tag{A.13}$$

Considering $\omega_3 \to \infty$ ($\gamma \to 0$), and use of the "Dirac delta function" $\delta(x)$

$$\lim_{\gamma \to 0} \frac{\gamma^2}{[x^2 + (1 - x^2)\, \gamma^2]^{3/2}} = 2\, \delta(x), \qquad \int \delta(x)\, f(x)\, dx = f(0) \tag{A.14}$$

yields, with $\omega_2 = 1$,

$$\mathbb{P} = \frac{1}{2\pi} \int_0^{2\pi} \mathbf{\Gamma}(\Theta = \frac{\pi}{2}, \Phi)\, \mathrm{d}\Phi \tag{A.15}$$

Next, we consider the numerator of Eqn. (A.10), $\mathbf{\Gamma} = \boldsymbol{\xi} \overset{s}{\otimes} \mathbf{K}^{-1} \overset{s}{\otimes} \boldsymbol{\xi}$. Expressing $\boldsymbol{\xi}$ and $\mathbf{K}$ in terms of the base vectors $\mathbf{w}_1$ and $\mathbf{w}_2$, while adopting $z = \cot\Phi$, yields

$$\boldsymbol{\xi} = \cos\Phi\, \mathbf{w}_1 + \sin\Phi\, \mathbf{w}_2 = \sin\Phi(z\, \mathbf{w}_1 + \mathbf{w}_2) \tag{A.16}$$

$$\begin{aligned} \mathbf{K} &= \boldsymbol{\xi} \cdot \mathbb{C} \cdot \boldsymbol{\xi} = \\ &= \sin^2\Phi\, ((z\, \mathbf{w}_1 + \mathbf{w}_2) \cdot \mathbb{C} \cdot (z\, \mathbf{w}_1 + \mathbf{w}_2))\, \sin^2\Phi\, \underbrace{(z^2\mathbf{Q} + z(\mathbf{R} + \mathbf{R}^T) + \mathbf{T})}_{\mathbf{K}'(z)} \end{aligned} \tag{A.17}$$

when having introduced the second-order tensors $\mathbf{Q}, \mathbf{R}$ and $\mathbf{T}$ as

$$\mathbf{Q} = \mathbf{w}_1 \cdot \mathbb{C} \cdot \mathbf{w}_1, \;\; \mathbf{R} = \mathbf{w}_1 \cdot \mathbb{C} \cdot \mathbf{w}_2, \;\; \mathbf{T} = \mathbf{w}_2 \cdot \mathbb{C} \cdot \mathbf{w}_2 \tag{A.18}$$

$$\mathbf{K}(\Phi) = \sin^2\Phi \, \underbrace{(z^2\mathbf{Q} + z(\mathbf{R} + \mathbf{R}^T) + \mathbf{T})}_{\mathbf{K}'(z)} \tag{A.19}$$

$\mathbf{K}'(z)$ is a second-order polynomial. In order to obtain the inverse of $\mathbf{K}'(z)$, we use the matrix of cofactors (algebraic complements) $\mathbf{co}\,\mathbf{K}'$,

$$(\mathbf{K}(z))^{-1} = \frac{1}{\sin^2\Phi}(\mathbf{K}')^{-1} = \frac{1}{\sin^2\Phi}\frac{1}{\det \mathbf{K}'}(\mathbf{co}\,\mathbf{K}') \tag{A.20}$$

The determinant of $\mathbf{K}'$, $\det \mathbf{K}'$, is a sixth-order polynomial. Thus

$$\begin{aligned} \mathbf{\Gamma} &= \boldsymbol{\xi} \overset{s}{\otimes} \mathbf{K}^{-1} \overset{s}{\otimes} \boldsymbol{\xi} = \frac{1}{\sin^2\Phi}\frac{1}{\det \mathbf{K}'}(\boldsymbol{\xi} \overset{s}{\otimes} (\mathbf{co}\,\mathbf{K}') \overset{s}{\otimes} \boldsymbol{\xi}) = \\ &= \frac{1}{\sin^2\Phi}\frac{1}{\det \mathbf{K}'}(\sin^2\Phi\,(z\,\mathbf{w}_1 + \mathbf{w}_2) \overset{s}{\otimes} (\mathbf{co}\,\mathbf{K}') \overset{s}{\otimes} (z\,\mathbf{w}_1 + \mathbf{w}_2)) \end{aligned} \tag{A.21}$$

Insertion of Eqn. (A.21) into Eqn. (A.15) and use of $\Phi = \operatorname{arccot} z$ yields

$$\mathbb{P} = \frac{1}{2\pi}\int_{\Phi=0}^{2\pi} \mathbf{\Gamma}\,\mathrm{d}\Phi = \frac{1}{2\pi}2\int_{z=-\infty}^{\infty} \mathbf{\Gamma}\frac{\mathrm{d}z}{1+z^2} \tag{A.22}$$

$$= \frac{1}{\pi}\int_{-\infty}^{\infty} \frac{(z\,\mathbf{w}_1 + \mathbf{w}_2) \overset{s}{\otimes} (\mathbf{co}\,\mathbf{K}') \overset{s}{\otimes} (z\,\mathbf{w}_1 + \mathbf{w}_2)}{(\det \mathbf{K}')\,(1+z^2)}\,\mathrm{d}z \tag{A.23}$$

The integrand in (A.23) is a rational fraction with a sixth-order polynomial in the numerator and an eighth-order polynomial in the denominator. Hence, the integration can be based on the Residue Theorem:

$$\int_{-\infty}^{\infty} f(z)\,dz = 2i\pi \sum_j \mathrm{Res}(f,\, z_j), \tag{A.24}$$

where z_j are the poles with a positive imaginary part, of the function $f(z)$.

Publication B

Micromechanics of crystal interfaces in polycrystalline solid phases of porous media: fundamentals and application to strength of hydroxyapatite biomaterials (Fritsch et al. 2007a)

Authored by Andreas Fritsch, Luc Dormieux, Christian Hellmich, and Julien Sanahuja
Published in *Journal of Materials Science*, Volume 42, pages 8824–8837

Interfaces are often believed to play a role in the mechanical behavior of mineralized biological and biomimetic materials. This motivates the micromechanical description of the elasticity and brittle failure of interfaces between crystals in a (dense) polycrystal, which serves as the skeleton of a porous material defined one observation scale above. Equilibrium and compatibility conditions, together with a suitable matrix-inclusion problem with a compliant interface, yield the homogenized elastic properties of the polycrystal, and of the porous material with polycrystalline solid phase. Incompressibility of single crystals guarantees finite shear stiffness of the polycrystal, even for vanishing interface stiffness, while increasing the latter generally leads to an increase of polycrystal shear stiffness. Corresponding elastic energy expressions give access to effective stresses representing the stress heterogeneities in the microstructures, which induce brittle failure. Thereby, Coulomb-type brittle failure of the crystalline interfaces implies Drucker-Prager-type (brittle, elastic limit-type) failure properties at the scale of the polycrystal. At the even higher scale of the porous material, high interfacial rigidities or low interfacial friction angles may result in closed elastic domains, indicating material failure even under hydrostatic pressure. This micromechanics model can satisfactorily reproduce the experimental strength data of different (brittle) hydroxyapatite biomaterials, across largely variable porosities. Thereby, the brittle failure criteria can be well approximated by micromechanically derived criteria referring to ductile solid matrices, both criteria being even identical if the solid matrix is incompressible.

B.1 Introduction

Interfaces are believed to often play a fundamental role in the mechanical behavior of hierarchically organized biological materials. Accordingly, much attention has been paid to the polymer-filled interfaces between ceramic tablets in nacre (Gennes and Okumura 2000; Okumura and Gennes 2001; Katti and Katti 2001; Katti et al. 2001; Okumura 2002, 2003; Barthelat et al. 2007), but the importance of interfacial behavior was also discussed for other classes of biological materials, such as bone (Tai et al. 2006).

To gain insight into these material systems, material/microstructure models have been developed within different theoretical frameworks, such as fracture mechanics and scaling laws (Gennes and Okumura 2000; Okumura and Gennes 2001; Okumura 2002, 2003), large-scale elastoplastic Finite Element analyses (Katti and Katti 2001; Katti et al. 2001; Tai et al. 2006), or periodic homogenization on the basis of a unit cell discretized by Finite Elements (Barthelat et al. 2007).

In addition to such periodic, FE-based ('computational') homogenization approaches, analytical and/or semianalytical approaches of random homogenization (continuum micromechanics (Zaoui 1997b, 2002)) have been recently used as to effectively predict the elastic properties of complicated hierarchically structured material systems (such as bone (Hellmich and Ulm 2002b; Hellmich et al. 2004b,a; Fritsch and Hellmich 2007), wood (Hofstetter et al. 2005, 2006), concrete (Bernard et al. 2003; Ulm et al. 2004; Hellmich and Mang 2005), or shale (Ulm et al. 2005)), from the elasticity and the mechanical interactions – over different observation scales – of nanoscaled elementary components. Thereby, not every single detail of the highly random microstructures, but only the essential morphological features are considered, in terms of homogeneous subdomains (material phases) inside representative volume elements (RVEs, Fig. B.1), their volume fractions, their elasticity, and their mechanical interaction. Theoretically, it has been recently well understood how to extend these homogenization techniques to the *ductile* failure of (*bulk*) phases (Dormieux and Maghous 2000; Bernaud et al. 2002; Barthélémy and Dormieux 2003, 2004; Dormieux et al. 2006c,a) (while applications to real materials (Lemarchand et al. 2002) are more rare than for the elastic case). In comparison, the treatment of *brittle* failure and of *interfaces* in the framework of random homogenization is still a very open field: It is the focus of this paper – both fundamentally, and in view of the failure of biomimetic hydroxyapatite biomaterials.

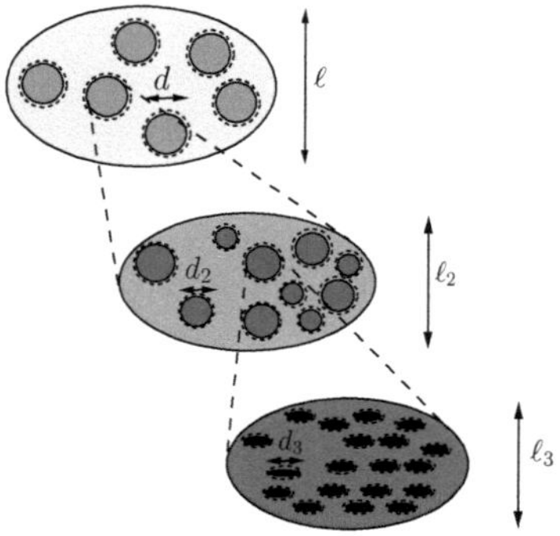

Figure B.1: Multistep homogenization: Properties of phases (with characteristic lengths of d and d_2, respectively) inside RVEs with characteristic lengths of ℓ or ℓ_2, respectively, are determined from homogenization over smaller RVEs with characteristic lengths of $\ell_2 \leq d$ and $\ell_3 \leq d_2$, respectively.

Extending very recent results (Sanahuja and Dormieux 2005; Dormieux et al. 2007), where inclusion coatings and interfaces in porous polycrystals were modeled, we here tackle the description of the elasticity and failure of interfaces between crystals in a (dense) polycrystal, which serves as the skeleton of a porous material defined one observation scale above (Fig. B.2). Thereby, we show characteristic features of a corresponding new micromechanics model, which is based on matrix-inclusion problems with compliant interfaces (Hashin 1991; Hervé and Zaoui 1993; Zhong and Meguid 1997), and which turns out to reasonably explain the behavior of porous hydroxyapatite biomaterials, especially for their brittle failure in the compressive regime.

A_C	surface area of spherical crystal with radius a
A_{ex}	constant in solution of matrix-inclusion problem with compliant interface
A_i	surface area of crystal i
A_{in}	constant in solution of matrix-inclusion problem with compliant interface
a	characteristic crystal radius
B_{ex}	constant in solution of matrix-inclusion problem with compliant interface
B_{in}	constant in solution of matrix-inclusion problem with compliant interface
C_{ex}	constant in solution of matrix-inclusion problem with compliant interface
$\mathbb{c}_C$	fourth-order stiffness tensor of single crystals within the RVE V_{poly}
$\mathbb{C}_{poly}$	fourth-order homogenized stiffness tensor of polycrystal with compliant interfaces
$\mathbb{C}_{PORO}$	fourth-order homogenized stiffness tensor of a porous material the solid phase of which is a polycrystal with weak interfaces
$\boldsymbol{E}_{poly}$	second-order 'macroscopic' strain tensor (related to RVE V_{poly} of polycrystal with compliant interfaces)
$\boldsymbol{E}_0$	uniform strain imposed at infinity of matrix surrounding inclusion with compliant interface
$E_{poly,v}$	'macroscopic' volumetric strain (related to RVE V_{poly} of polycrystal with compliant interfaces)
$E_{poly,d}$	'macroscopic' equivalent deviatoric strain (related to RVE V_{poly} of polycrystal with compliant interfaces)
$\underline{e}_r$	radial unit vector
$\underline{e}_1, \underline{e}_2, \underline{e}_3$	unit base vectors of Cartesian base frame
f_i	volume fraction of crystal i within the RVE V_{poly}
h	cohesion of interfaces between single crystals
$\mathbb{I}$	fourth-order identity tensor
$\mathcal{I}$	entity of interfaces within polycrystalline RVE V_{poly}
$\mathcal{I}_{ij}$	interface between crystals i and j
$\mathbb{J}$	volumetric part of fourth-order identity tensor $\mathbb{I}$
$\mathbb{K}$	deviatoric part of fourth-order identity tensor $\mathbb{I}$
$\boldsymbol{K}$	second-order interface stiffness tensor
$\boldsymbol{K}' = 2\boldsymbol{K}$	second-order interface stiffness tensor in matrix-inclusion problem with compliant interface
K_n	normal interface stiffness (component of $\boldsymbol{K}$)
K_t	tangential interface stiffness (component of $\boldsymbol{K}$)
k_C	bulk modulus of single crystals
k_{poly}	homogenized bulk modulus of polycrystal with compliant interfaces (RVE V_{poly})
k_{PORO}	homogenized bulk modulus of a porous material the solid phase of which is a polycrystal with compliant interfaces
$\underline{n}$	normal vector onto surface of a single crystal
RVE	representative volume element
r	radial coordinate in spherical coordinate system
$\mathbb{S}$	fourth-order Eshelby tensor for spherical inclusions
$\underline{T}$	traction force vector acting on surface element of interface
T_n	normal component of $\underline{T}$
T_t	tangential component of $\underline{T}$
T_t^{cr}	critical (maximum) tangential traction bearable by intercrystalline interface

$\underline{t}$	tangential vector to surface of a single crystal
tr	trace of a second-order tensor
V_C	volume of spherical crystal with radius a
∂V_C	surface of spherical crystal with radius a
V_i	volume of crystal i
∂V_i	surface of crystal i
V_{poly}	volume of an RVE of polycrystal with compliant interfaces
V_{PORO}	volume of an RVE of porous material the solid phase of which is a polycrystal with compliant interfaces
V_S	volume of solid phase within the RVE V_{PORO}
$\underline{x}$	position vector within an RVE, either V_{poly} or V_{PORO}
α	friction angle of interfaces between single crystals
δ_{ij}	Kronecker delta (components of second-order identity tensor **1**)
$\delta_{\mathcal{I}}$	Dirac distribution supported on $\mathcal{I}$
ε	second-order strain tensor field within single crystals filling RVE V_{poly} of polycrystal with compliant interfaces
θ	latitudinal coordinate of spherical coordinate system
$\kappa = \frac{K_t' a}{\mu_C}$	dimensionless quantity related to rigidity of interface
μ_C	shear modulus of single crystals
μ_{poly}	homogenized shear modulus of polycrystal with compliant interfaces (RVE V_{poly})
μ_{PORO}	homogenized shear modulus of a porous material the solid phase of which is a polycrystal with compliant interfaces
ν_{poly}	homogenized Poisson's ratio of polycrystal with compliant interfaces (RVE V_{poly})
$\underline{\xi}$	displacements within and at the boundary of RVE V_{poly}
$[\![\underline{\xi}]\!]$	displacement discontinuity at the interfaces between crystals
$[\![\xi_n]\!]$	normal component of $[\![\underline{\xi}]\!]$
$[\![\xi_t]\!]$	tangential component of $[\![\underline{\xi}]\!]$
$[\underline{\xi}]$	displacement discontinuity at compliant interface of 'generalized' matrix-inclusion problem
$\underline{\xi}_i$, $\underline{\xi}_j$	displacements along interface $\mathcal{I}_{ij}$, in crystal i and j, respectively
$\bar{\underline{\xi}}$	mean displacement at the interface $\mathcal{I}_{ij}$
$\underline{\xi}_{in}$	displacement field inside the inclusion surrounded by compliant interface and infinite matrix (related to 'generalized' matrix-inclusion problem)
$\underline{\xi}_{ex}$	displacement field throughout the matrix surrounding inclusion coated by compliant interface (related to 'generalized' matrix-inclusion problem)
$\mathbf{\Sigma}_{poly}$	second-order 'macroscopic' stress tensor (related to RVE V_{poly} of polycrystal with weak interfaces)
$\Sigma_{poly,m}$	'macroscopic' mean stress (related to RVE V_{poly} of polycrystal with weak interfaces)
$\Sigma_{poly,d}$	'macroscopic' equivalent deviatoric stress (related to RVE V_{poly} of polycrystal with weak interfaces)
$\mathbf{\Sigma}_{PORO}$	second-order macroscopic stress tensor (related to RVE V_{PORO} of porous material the solid phase of which is a polycrystal with weak interfaces)
$\Sigma_{PORO,m}$	macroscopic mean stress (related to RVE V_{PORO} of porous material the solid phase of which is a polycrystal with weak interfaces)

$\Sigma_{PORO,d}$	macroscopic equivalent deviatoric stress (related to RVE V_{PORO} of porous material the solid phase of which is a polycrystal with weak interfaces)
$\boldsymbol{\sigma}$	second-order stress tensor field within single crystals filling RVE V_{poly} of polycrystal with compliant interfaces
$\boldsymbol{\sigma}_{in}$	stress field inside the inclusion surrounded by compliant interface and infinite matrix (related to 'generalized' matrix-inclusion problem)
$\boldsymbol{\sigma}_{ex}$	stress field throughout the matrix surrounding inclusion coated by compliant interface (related to 'generalized' matrix-inclusion problem)
ϕ	longitudinal coordinate of spherical coordinate system
φ	volume fraction of pores within the RVE V_{PORO}
$\chi = \frac{\mu_C}{k_C}$	dimensionless quantity related to compressibility of single crystals
Ψ	macroscopic energy density
$\mathbf{1}$	second-order identity tensor
$\cdot$	first-order tensor contraction
$:$	second-order tensor contraction
$\otimes$	dyadic product of tensors

Table B.1: List of symbols.

B.2 Fundamentals of continuum micromechanics – representative volume element

In continuum micromechanics (Hill 1963; Suquet 1997a; Zaoui 1997b, 2002), a material is understood as a macro-homogeneous, but micro-heterogeneous body filling a representative volume element (RVE) with characteristic length ℓ, $\ell \gg d$, d standing for the characteristic length of inhomogeneities within the RVE (see Fig. B.1), and $\ell \ll \mathcal{L}$, $\mathcal{L}$ standing for the characteristic lengths of geometry or loading of a structure built up by the material defined on the RVE. In general, the microstructure within one RVE is so complicated that it cannot be described in complete detail. Therefore, quasi-homogeneous subdomains with known physical quantities (such as volume fractions or elastic properties) are reasonably chosen. They typically include 3D subdomains, and may also include the 2D interfaces between the 3D subdomains. They are called material phases; bulk and interface phases, respectively. The 'homogenized' mechanical behavior of the overall material, i.e. the relation between homogeneous deformations acting on the boundary of the RVE and resulting (average) stresses, or the ultimate stresses sustainable by the RVE, can then be estimated from the mechanical behavior of the aforementioned homogeneous phases (representing the inhomogeneities within the RVE), their dosages within the RVE, their characteristic shapes, and their interactions. If a single phase exhibits a heterogeneous microstructure itself, its mechanical behavior can be estimated by introduction of an RVE within this phase, with dimensions $\ell_2 \leq d$, comprising again smaller phases with characteristic length $d_2 \ll \ell_2$, and so on, leading to a multistep homogenization scheme (see Fig. B.1).

B.3 Micromechanics of polycrystal with weak interfaces

B.3.1 Micromechanical representation

We consider an RVE with volume V_{poly} [Fig. B.2(a) and Fig. B.9(a)], hosting single crystals of typically quasi-spherical shape and of volume V_i, separated from each other by very thin (essentially 2D) interfaces $\mathcal{I}_{ij}$ between crystals i and j, all interfaces making up the entity of interfaces $\mathcal{I}$, $\cup \mathcal{I}_{ij} = \mathcal{I}$, see Fig. B.2. 'Macroscopic' strains $\boldsymbol{E}_{poly}$ are imposed at the boundary of the RVE V_{poly} in terms of displacements $\underline{\xi}$,

$$\text{on}\, \partial V_{poly}: \quad \underline{\xi}(\underline{x}) = \boldsymbol{E}_{poly} \cdot \underline{x} \tag{B.1}$$

with $\underline{x}$ as the position vector within the RVE. The geometrical compatibility of (B.1) with the local 'microscopic' strains $\boldsymbol{\varepsilon}(\underline{x})$ in the crystals and the displacement discontinuities $[\![\underline{\xi}]\!] = \underline{\xi}_j - \underline{\xi}_i$

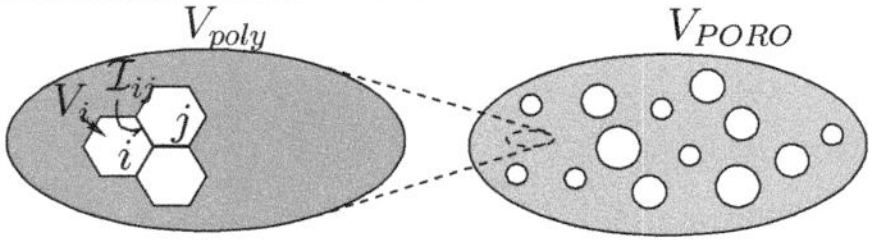

(a) (b)

Figure B.2: (a) Polycrystal with interfaces (schematic representation of volume V_i of crystal i and interface $\mathcal{I}_{ij}$ between crystals i and j), serving as skeleton in a porous material at larger observation scale (b).

at the interfaces $\mathcal{I}_{ij}$ between the crystals i and j implies (Dormieux et al. 2007)

$$\boldsymbol{E}_{poly} = \frac{1}{V_{poly}} \left(\int_{V_{poly}} \boldsymbol{\varepsilon}(\underline{x}) \,\mathrm{d}V + \sum_{ij} \int_{\mathcal{I}_{ij}} [\![\underline{\xi}]\!] \overset{s}{\otimes} \underline{n} \,\mathrm{d}S \right) =$$

$$= \frac{1}{V_{poly}} \sum_i \int_{\partial V_i} \bar{\underline{\xi}} \overset{s}{\otimes} \underline{n} \,\mathrm{d}S = \sum_i \frac{f_i}{V_i} \int_{\partial V_i} \bar{\underline{\xi}} \overset{s}{\otimes} \underline{n} \,\mathrm{d}S \tag{B.2}$$

with location vector $\underline{x}$, normal $\underline{n}$ onto the spherical surface of the crystals,

$$\bar{\underline{\xi}} = (\underline{\xi}_i + \underline{\xi}_j)/2 = \underline{\xi}_j - [\![\underline{\xi}]\!]/2 = \underline{\xi}_i + [\![\underline{\xi}]\!]/2 \tag{B.3}$$

as the mean displacement at the interface $\mathcal{I}_{ij}$, V_i and $f_i = V_i/V_{poly}$ as the volume and the volume fraction of the i-th crystal, and ∂V_i as its surface with area A_i. For crystals of the same shape and size (with volume V_C and surface ∂V_C), and indiscernible average mean displacements at their surfaces, (B.2) can be transformed to

$$\boldsymbol{E}_{poly} = \frac{1}{V_C} \int_{\partial V_C} \bar{\underline{\xi}} \overset{s}{\otimes} \underline{n} \,\mathrm{d}S \tag{B.4}$$

The corresponding 'macroscopic' stresses $\boldsymbol{\Sigma}_{poly}$ are equal to the spatial average of the (equilibrated) local stresses $\boldsymbol{\sigma}(\underline{x})$ inside the RVE V_{poly},

$$\boldsymbol{\Sigma}_{poly} = \langle \boldsymbol{\sigma}(\underline{x}) \rangle = \frac{1}{V_{poly}} \int_{V_{poly}} \boldsymbol{\sigma}(\underline{x}) \,\mathrm{d}V =$$

$$= \sum_i \frac{f_i}{V_i} \int_{V_i} \boldsymbol{\sigma}(\underline{x}) \,\mathrm{d}V =$$

$$= \sum_i \frac{f_i}{V_i} \int_{\partial V_i} \underline{x} \otimes [\boldsymbol{\sigma}(\underline{x}) \cdot \underline{n}(\underline{x})] \,\mathrm{d}S \tag{B.5}$$

For spherical crystals with radius a, surface ∂V_C with area $A_C = 4\pi a^2$, and volume $V_C = 4/3\pi a^3$, (B.5) can be further transformed,

$$\begin{aligned}\boldsymbol{\Sigma}_{poly} &= \sum_i \frac{f_i}{\frac{4}{3}\pi a^3} \int_{\partial V_C} a\, \underline{e}_r(\underline{x}) \otimes [\boldsymbol{\sigma}(\underline{x}) \cdot \underline{e}_r(\underline{x})]\, \mathrm{d}S = \\ &= \sum_i \frac{3 f_i}{A_C} \int_{\partial V_C} \underline{e}_r(\underline{x}) \otimes [\boldsymbol{\sigma}(\underline{x}) \cdot \underline{e}_r(\underline{x})]\, \mathrm{d}S = \\ &= \frac{1}{V_C} \int_{\partial V_C} a\, \underline{n}(\underline{x}) \otimes [\boldsymbol{\sigma}(\underline{x}) \cdot \underline{n}(\underline{x})]\, \mathrm{d}S = \\ &= \frac{3}{A_C} \int_{\partial V_C} \underline{n}(\underline{x}) \otimes [\boldsymbol{\sigma}(\underline{x}) \cdot \underline{n}(\underline{x})]\, \mathrm{d}S = \\ &= \frac{1}{V_C} \int_{V_C} \boldsymbol{\sigma}(\underline{x})\, \mathrm{d}V \end{aligned} \tag{B.6}$$

with radial unit vector $\underline{e}_r$ being identical to the normal $\underline{n}$. Since the microscopic stresses are equilibrated ($\mathrm{div}\, \boldsymbol{\sigma} = \underline{0}$), (B.5) and (B.6) imply (Hill 1963), (Dormieux 2005, p. 118), that the 'macroscopic' stresses act as traction forces $\boldsymbol{\Sigma}_{poly} \cdot \underline{n}$ both at the boundary of the RVE, ∂V_{poly}, and those of single crystals, ∂V_C,

$$\text{on}\, \partial V_{poly}\, \text{and}\, \partial V_C : \quad \boldsymbol{\sigma}(\underline{x}) \cdot \underline{n}(\underline{x}) = \boldsymbol{\Sigma}_{poly} \cdot \underline{n}(\underline{x}) \tag{B.7}$$

The relation between $\boldsymbol{\Sigma}_{poly}$ and $\boldsymbol{E}_{poly}$ depends on the constitutive behavior of the single crystals and of the interfaces between them.

B.3.2 Constitutive behavior of interfaces and single crystals

The interfaces are the weakest locations of the material, the load bearing capacities of which are bounded according to a Coulomb-type law,

$$\forall \underline{x} \in \mathcal{I}_{ij} : \; T_t(\underline{x}) \leq T_t^{cr} = \alpha(h - T_n(\underline{x})) \tag{B.8}$$

with friction angle α, cohesion h, and T_t and T_n as the tangential and normal components of the traction force $\underline{T} = T_n\, \underline{n} + T_t\, \underline{t}$ acting on an infinitesimal interface area around $\underline{x}$, with normal $\underline{n}$, and $\underline{t}$ as the tangential unit vector, $\underline{t} \cdot \underline{n} = 0$. We consider brittle interface failure once a critical value $T_t = T_t^{cr}$ is reached in (B.8).

Below this critical value, the interface behaves linear elastically, i.e. the interface traction $\underline{T}(\underline{x})$ is related to a displacement discontinuity $[\![\underline{\xi}]\!](\underline{x})$ encountered when crossing the interface $\mathcal{I}_{ij}$ along $\underline{n}(\underline{x})$:

$$\underline{T}(\underline{x}) = \boldsymbol{K} \cdot [\![\underline{\xi}]\!](\underline{x})$$

with

$$\boldsymbol{K} = K_n\, \underline{n} \otimes \underline{n} + K_t\, (\mathbf{1} - \underline{n} \otimes \underline{n}), \; K_n \to \infty \tag{B.9}$$

$\boldsymbol{K}$ is the second-order interface stiffness tensor with infinite normal component K_n (no mutual interpenetration of crystals), and positive tangential component K_t (allowing for relative tangential movements of crystal surfaces). Also the bulk crystal phase inside the RVE V_{poly} behaves linear elastically,

$$\forall \underline{x} \in V_i : \quad \boldsymbol{\sigma}(\underline{x}) = \mathbb{c}_C : \boldsymbol{\varepsilon}(\underline{x}) \tag{B.10}$$

with $\mathbb{c}_C = 3k_C\,\mathbb{J} + 2\mu_C\,\mathbb{K}$ as the isotropic elastic stiffness of the bulk material phase comprising all single crystals; with bulk modulus k_C and shear modulus μ_C. $\mathbb{J} = 1/3\,\mathbf{1}\otimes\mathbf{1}$ and $\mathbb{K} = \mathbb{I} - \mathbb{J}$ are the volumetric and the deviatoric part of the fourth-order identity tensor $\mathbb{I}$, with components $I_{ijkl} = 1/2(\delta_{ik}\delta_{jl} + \delta_{il}\delta_{kj})$; the components of the second-order unit tensor $\mathbf{1}$, δ_{ij} (Kronecker delta), read as $\delta_{ij} = 1$ for $i = j$ and $\delta_{ij} = 0$ for $i \neq j$.

The assumption of crystal isotropy deserves to be commented, since single crystals are generally anisotropic, including approximately transversely isotropic hydroxayapatite (Katz and Ukraincik 1971). However, hydroxyapatite anisotropy is not very pronounced (Katz and Ukraincik 1971), and in addition, the disorder of crystals (and of their principal material directions) probably renders isotropic phase properties as suitable approximation for the purpose of polycrystal property homogenization. This was recently shown quantitatively for polycrystals consisting of perfectly disordered needles, being either isotropic or anisotropic (Fritsch et al. 2006).

B.3.3 Homogenized elasticity of polycrystal with compliant interfaces

As long as the interfaces behave elastically, the relation between $\boldsymbol{\Sigma}_{poly}$ and $\boldsymbol{E}_{poly}$ reads as

$$\boldsymbol{\Sigma}_{poly} = \mathbb{C}_{poly} : \boldsymbol{E}_{poly} \tag{B.11}$$

with the 'macroscopic' homogenized stiffness tensor of the polycrystal, $\mathbb{C}_{poly} = 3k_{poly}\,\mathbb{J} + 2\mu_{poly}\,\mathbb{K}$, with bulk modulus k_{poly} and shear modulus μ_{poly}; depending on the local elastic properties $\mathbb{c}_C$ and K_t.

Following (Dormieux et al. 2007), the establishment of this dependence is based on the behavior of a composite solid consisting of a spherical inclusion of radius a and a compliant interface coating the inclusion, being itself embedded in an infinite matrix exhibiting the elastic properties $\mathbb{C}_{poly}$ of the homogenized polycrystal, and being subjected to uniform strains $\boldsymbol{E}_0$ at infinity (Fig. B.3). Mathematically, we have

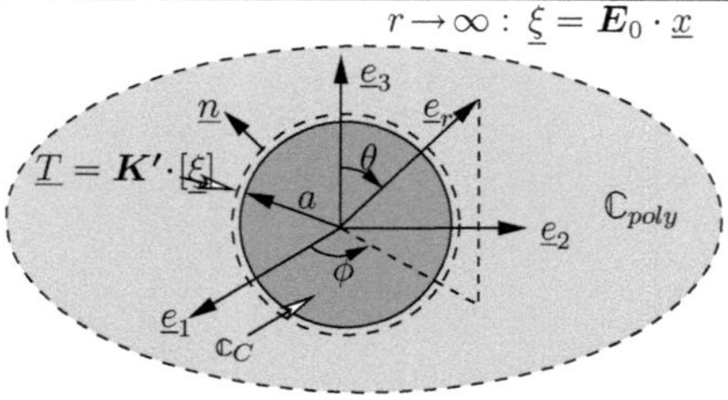

Figure B.3: Matrix-inclusion problem with compliant interface ('generalized Eshelby problem'): A spherical inclusion with interface is embedded in an infinite matrix subjected to uniform strain $\boldsymbol{E}_0$ at infinity. The elastic properties of the matrix are those of the homogenized material.

$$\begin{aligned}
r < a \quad &: \quad \boldsymbol{\sigma} = \mathbb{C}_C : \boldsymbol{\varepsilon} \\
r = a \quad &: \quad \underline{T} = \boldsymbol{K}' \cdot [\underline{\xi}] \\
&\quad\;\; \text{with } [\underline{\xi}] = [\![\underline{\xi}]\!]/2,\; \boldsymbol{K}' = 2\boldsymbol{K} \\
r > a \quad &: \quad \boldsymbol{\sigma} = \mathbb{C}_{poly} : \boldsymbol{\varepsilon} \\
r \to \infty \quad &: \quad \underline{\xi} \to \boldsymbol{E}_0 \cdot \underline{x}
\end{aligned} \tag{B.12}$$

For determination of k_{poly}, a purely spherical deformation, $\boldsymbol{E}_0 = E_0\,\mathbf{1}$ is imposed at $r \to \infty$. Spherical symmetry of both the loading and the geometry of the considered solid implies vanishing tangential displacement discontinuities at the inclusion interface, $[\![\xi_t]\!] \equiv 0$. Since $K_n \to \infty$, also $[\![\xi_n]\!] = 0$ (no mutual interpenetration of crystals), and the matrix inclusion problem with compliant interfaces reduces to the classical Eshelby-type inclusion problem with a perfect, rigid interface (Eshelby 1957). Then, consideration of only one bulk phase (the crystals) implies that the overall bulk modulus k_{poly} is identical to the crystal bulk modulus k_C,

$$k_{poly} \equiv k_C \tag{B.13}$$

For determination of μ_{poly}, a purely deviatoric defomation, $\boldsymbol{E}_0 = E_0(\underline{e}_1 \otimes \underline{e}_1 - \underline{e}_3 \otimes \underline{e}_3)$, is imposed (see Fig. B.3 for the Cartesian base frame $\underline{e}_1$, $\underline{e}_2$, $\underline{e}_3$). The mathematical form of the displacement field in the exterior region, $r > a$ (the homogenized material), $\underline{\xi}_{ex}$, is established in the line of (Hervé and Zaoui 1993), and reads in spherical coordinates (see Fig. B.3 for

Eulerian angles ϕ and θ) as

$$\frac{\xi_{ex,r}}{E_0} = (A_{ex}\, r + 3\frac{B_{ex}}{r^4} + \frac{5-4\nu_{poly}}{1-2\nu_{poly}}\frac{C_{ex}}{r^2})(\cos^2\phi\sin^2\theta - \cos^2\theta)$$

$$\frac{\xi_{ex,\theta}}{E_0} = \frac{1}{2}(A_{ex}\, r - 2\frac{B_{ex}}{r^4} + 2\frac{C_{ex}}{r^2})\sin 2\theta(1+\cos^2\phi)$$

$$\frac{\xi_{ex,\phi}}{E_0} = -\frac{1}{2}(A_{ex}\, r - 2\frac{B_{ex}}{r^4} + 2\frac{C_{ex}}{r^2})\sin\theta\sin 2\phi \tag{B.14}$$

where ν_{poly} is the Poisson's ratio of the polycrystal with weak interfaces,

$$\nu_{poly} = \frac{3k_{poly} - 2\mu_{poly}}{6k_{poly} + 2\mu_{poly}} \tag{B.15}$$

The boundary condition in $(B.12)_4$ directly implies $A_{ex} = 1$, while the constants B_{ex} and C_{ex} will follow from interface conditions.

Inside the inclusion ($r < a$, the solid crystal phase), the displacement field $\underline{\xi}_{in}$ reads as

$$\frac{\xi_{in,r}}{E_0} = (A_{in}\, r + B_{in}\, r^3)\,(\cos^2\phi\sin^2\theta - \cos^2\theta)$$

$$\frac{\xi_{in,\theta}}{E_0} = \frac{1}{2}(A_{in}\, r + \frac{(11\mu_C + 15\, k_C)B_{in}r^3}{3\,(3k_C - 2\mu_C)})\sin 2\theta(1+\cos^2\phi)$$

$$\frac{\xi_{in,\phi}}{E_0} = -\frac{1}{2}(A_{in}\, r + \frac{(11\mu_C + 15\, k_C)B_{in}r^3}{3\,(3k_C - 2\mu_C)})\sin\theta\sin 2\phi \tag{B.16}$$

The four remaining constants B_{ex}, C_{ex}, A_{in} and B_{in} are determined by enforcing equilibrium of forces at the interface $r = a$:

$$\underline{T} = \boldsymbol{\sigma}_{in}\cdot\underline{n} = \boldsymbol{\sigma}_{ex}\cdot\underline{n} = \boldsymbol{K}'\cdot[\underline{\xi}] \tag{B.17}$$

together with constitutive laws $(B.12)_1$, $(B.12)_2$ and $(B.12)_3$, see Appendix B.6. This solution for the displacement fields $\underline{\xi}_{in}$ and $\underline{\xi}_{ex}$ gives access to the traction forces at the interfaces $\underline{T}(r=a) = \boldsymbol{\sigma}\cdot\underline{n}(r=a) = \boldsymbol{K'}\cdot[\underline{\xi}_{ex}(r=a^+) - \underline{\xi}_{in}(r=a^-)]$. Their use for estimating the traction forces at the interfaces within the polycrystalline RVE V_{poly} yields the corresponding 'macroscopic' stress $\boldsymbol{\Sigma}_{poly}$ according to (B.6) as

$$\boldsymbol{\Sigma}_{poly} = \frac{1}{V_C}\int_{\partial V_C} a\,\underline{n}\otimes(\boldsymbol{\sigma}\cdot\underline{n})(r=a)\,\mathrm{d}S \tag{B.18}$$

The solution for the displacements at $r = a^+$ turns out to be, according to $(B.12)_2$ and (B.3), a suitable estimate for the mean displacement $\bar{\underline{\xi}}$ at the crystal interface $\mathcal{I}_{ij}$. Use of this quantity in (B.4) yields the corresponding 'macroscopic' strains $\boldsymbol{E}_{poly}$ in the form

$$\boldsymbol{E}_{poly} = \frac{1}{V_C}\int_{\partial V_C} \underline{\xi}_{ex}(a^+)\overset{s}{\otimes}\underline{n}\,\mathrm{d}S \tag{B.19}$$

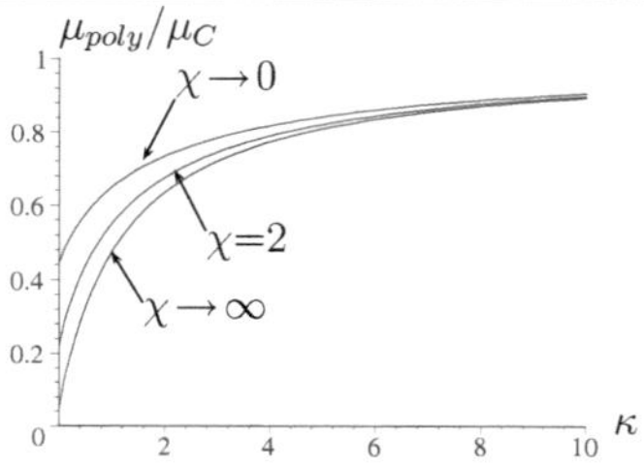

Figure B.4: Homogenized shear modulus μ_{poly} of polycrystal, as function of dimensionless quantity $\kappa = K_t' a/\mu_C$ (interfacial rigidity), for different crystal compressibilities $\chi = \mu_C/k_C$, Eq.(B.20).

Shear components $\Sigma_{poly,12}$ and $E_{poly,12}$ of 'macroscopic' stresses (B.18) and strains (B.19), together with (B.14)–(B.17) and (B.50)–(B.54), give access to μ_{poly}, via $\mu_{poly} = \Sigma_{12}/(2E_{12})$, yielding (after elimination of E_0) the following expression,

$$\frac{\mu_C}{\mu_{poly}} = 1 + 3\left[\frac{5\kappa}{2} + \left(\frac{\mu_C}{8\mu_{poly}} + \frac{6k_C + 17\mu_C}{57k_C + 4\mu_C}\right)^{-1}\right]^{-1} \tag{B.20}$$

with the dimensionless quantity $\kappa = K_t' a/\mu_C$. $\kappa\to\infty$ relates to a rigid interface. The higher the rigidity κ of the interface, the higher the overall polycrystal shear modulus (Fig. B.4), for different (dimensionless) compressibilities $\chi = \mu_C/k_C$ of the single crystals. Thereby, crystal incompressibility ($\chi \to 0$) guarantees finite overall shear stiffness even for an interface with vanishing stiffness ($\kappa = 0$), while a polycrystal built up of crystals with zero bulk modulus ($\chi\to\infty$) and connected through zero-stiffness interfaces ($\kappa = 0$) does not exhibit any shear stiffness (Fig. B.4), but still the bulk stiffness of the single crystals according to (B.13). In case of an incompressible solid ($k_C\to\infty$, $\chi = \mu_C/k_C\to 0$), it follows from (B.13) that $k_{poly}\to\infty$, and (B.20) reduces to

$$48(5+\kappa)\left(\frac{\mu_{poly}}{\mu_C}\right)^2 + (-114+9\kappa)\frac{\mu_{poly}}{\mu_C} - 57\kappa = 0 \tag{B.21}$$

B.3.4 Upscaled failure properties of polycrystal with weak interfaces

In order to determine the effective failure properties resulting from local failure characteristics (B.8) and from the interactions between interfaces and bulk single crystals, we are left with relating the local interface forces $\underline{T}(\underline{x}) \in \mathcal{I}$ to the 'macroscopic' stresses $\mathbf{\Sigma}_{poly}$, see (B.5). The tangential and normal traction forces, T_t and T_n, occuring in the interface failure criterion (B.8), are non-homogeneously distributed across the interfaces. Failure will occur where relatively high tangential traction forces encounter a relative low resistance due to relatively low normal traction forces. Instead of trying to model the actual force fields across the interfaces, we estimate the effect of the actual force distribution through so-called *effective* traction forces, as it is commonly done for stress, strain, or force fields in the context of continuum micromechanics (Suquet 1997a; Dormieux et al. 2007). In this line, we represent the failure-inducing interplay between moderate normal traction forces and tangential traction force *peaks* by means of two different *effective* measures for the normal and the tangential traction forces, respectively: (i) first-order moments of normal forces, and (ii) second-order moments of tangential forces.

The first-order moment of the normal traction forces, $\langle T_n \rangle$, is related to the 'macroscopic' mean stress $\Sigma_{poly,m}$ through

$$\Sigma_{poly,m} = \frac{1}{3}\mathrm{tr}\,\boldsymbol{\Sigma}_{poly} = \frac{1}{3}\mathrm{tr}\left(\frac{3}{A_C}\int_{\partial V_C} \underline{n}(\underline{x}) \otimes [\boldsymbol{\sigma}(\underline{x}) \cdot \underline{n}(\underline{x})]\,\mathrm{d}S\right) =$$

$$= \frac{1}{A_C}\int_{\partial V_C} \sigma_{rr}(\underline{x})\,\mathrm{d}S = \frac{1}{A_C}\int_{\partial V_C} T_n(\underline{x})\,\mathrm{d}S = \langle T_n \rangle \qquad \text{(B.22)}$$

(B.22) establishes a first link between the 'macroscopic' stress $\boldsymbol{\Sigma}_{poly}$ and the interface tractions $\underline{T}(\underline{x})$: We use this average (or first-order moment) of normal traction forces as to estimate the 'average' interface resistance T_t^{cr} in (B.8), according to

$$T_t^{cr} \approx \alpha(h - \langle T_n \rangle) \qquad \text{(B.23)}$$

However, use of the average tangential traction force $\langle T_t \rangle$ in failure criterion (B.8) is problematic since force peaks initializing failure may be cancelled out in the averaging process. As a remedy, we use the second-order moment $\sqrt{\langle T_t^2 \rangle}$ (also called quadratic average) as a characteristic or *effective* value for $T_t(\underline{x})$, in the line of (Kreher 1990; Dormieux et al. 2002, 2007). The relation between $\sqrt{\langle T_t^2 \rangle}$ and $\boldsymbol{\Sigma}_{poly}$ is established through energy considerations: The energy stored in the RVE V_{poly} can be expressed through the global 'macroscopic' energy density Ψ as

$$V_{poly}\,\Psi = \frac{1}{2} V_{poly}\,\boldsymbol{\Sigma}_{poly} : \boldsymbol{E}_{poly} =$$

$$= \frac{1}{2} V_{poly}\,\boldsymbol{E}_{poly} : \mathbb{C}_{poly} : \boldsymbol{E}_{poly} =$$

$$= V_{poly}\left(\frac{1}{2} k_{poly} E_{poly,v}^2 + 2\mu_{poly} E_{poly,d}^2\right) \qquad \text{(B.24)}$$

with 'macroscopic' volumetric strain $E_{poly,v} = \mathrm{tr}\,\boldsymbol{E}_{poly}$ and equivalent deviatoric strain $E_{poly,d} = \sqrt{1/2\,\boldsymbol{E}_{poly,d} : \boldsymbol{E}_{poly,d}}$, $\boldsymbol{E}_{poly,d} = \boldsymbol{E}_{poly} - 1/3 E_{poly,v}\mathbf{1}$.

In order to express Ψ from a microstructural viewpoint, we consider the local constitutive behavior of the interface [Eq.(B.9)] and of the bulk phase [Eq.(B.10)]. The corresponding 'macroscopic' elastic energy stored in the RVE reads as

$$V_{poly}\,\Psi = \frac{1}{2}\int_{V_{poly}} \boldsymbol{\sigma} : \boldsymbol{\varepsilon}\,\mathrm{d}V + \frac{1}{2}\int_{\mathcal{I}} \underline{T}\cdot [\![\underline{\xi}]\!]\,\mathrm{d}S =$$

$$= \frac{1}{2}\int_{V_{poly}} \boldsymbol{\varepsilon} : \mathbb{c}_C : \boldsymbol{\varepsilon}\,\mathrm{d}V + \frac{1}{2}\int_{\mathcal{I}} [\![\underline{\xi}]\!]\cdot \boldsymbol{K}\cdot [\![\underline{\xi}]\!]\,\mathrm{d}S \tag{B.25}$$

In order to extract $\langle T_t^2\rangle = \frac{1}{A_C}\int_{\mathcal{I}} T_t^2\,\mathrm{d}S$ from (B.25), variations of Ψ with varying K_t (holding merely $\boldsymbol{E}_{poly}$ fixed) are studied,

$$V_{poly}\frac{\partial\Psi}{\partial K_t} = \int_{V_{poly}} \frac{\partial\boldsymbol{\varepsilon}}{\partial K_t} : \boldsymbol{\sigma}\,\mathrm{d}V + \int_{\mathcal{I}} \frac{\partial[\![\underline{\xi}]\!]}{\partial K_t}\cdot \underline{T}\,\mathrm{d}S +$$

$$+\frac{1}{2}\int_{\mathcal{I}} [\![\underline{\xi}]\!]\cdot(\mathbf{1}-\underline{n}\otimes\underline{n})\cdot[\![\underline{\xi}]\!]\,\mathrm{d}S = \int_{V_{poly}} \frac{\partial\boldsymbol{\varepsilon}}{\partial K_t} : \boldsymbol{\sigma}\,\mathrm{d}V +$$

$$+\int_{V_{poly}} \frac{\partial}{\partial K_t}([\![\underline{\xi}]\!]\otimes\underline{n}\,\delta_{\mathcal{I}}) : \boldsymbol{\sigma}\,\mathrm{d}V + \frac{1}{2}\int_{\mathcal{I}} [\![\xi_t]\!]^2\,\mathrm{d}S \tag{B.26}$$

where $\underline{T} = \boldsymbol{\sigma}\cdot\underline{n}$ was considered and where $\delta_{\mathcal{I}}$ is the 'Dirac distribution' of support $\mathcal{I}$, $\int_V \delta_{\mathcal{I}} f\,\mathrm{d}V = \int_{\mathcal{I}} f\,\mathrm{d}S$. For transformation of (B.26), we extend Hill's lemma (Hill 1963) to the case of displacement discontinuities at the interfaces (Dormieux et al. 2007). Considering (B.5) and the format (2) for the 'macroscopic' strains $\boldsymbol{E}_{poly}$, (B.26) can be transformed to

$$V_{poly}\frac{\partial\Psi}{\partial K_t} = \int_{V_{poly}} \frac{\partial}{\partial K_t}(\boldsymbol{\varepsilon} + [\![\underline{\xi}]\!]\otimes\underline{n}\,\delta_{\mathcal{I}}) : \boldsymbol{\sigma}\,\mathrm{d}V +$$

$$+\frac{1}{2}\int_{\mathcal{I}} [\![\xi_t]\!]^2\,\mathrm{d}S = \frac{\partial\boldsymbol{E}_{poly}}{\partial K_t} : \boldsymbol{\Sigma}_{poly} + \frac{1}{2}\int_{\mathcal{I}} [\![\xi_t]\!]^2\,\mathrm{d}S \tag{B.27}$$

Fixed 'macroscopic' strains $\boldsymbol{E}_{poly}$ according to (B.1) imply $\partial\boldsymbol{E}_{poly}/\partial K_t = \mathbf{0}$, so that (B.27) becomes

$$V_{poly}\frac{\partial\Psi}{\partial K_t} = \frac{1}{2}\int_{\mathcal{I}} [\![\xi_t]\!]^2\,\mathrm{d}S = \frac{\mathcal{I}}{2}\left\langle [\![\xi_t]\!]^2\right\rangle \tag{B.28}$$

Identification of (B.28) with the derivation of the 'macroscopic' expression for the energy density (B.24) with respect to K_t yields

$$\frac{\mathcal{I}}{V_{poly}}\left\langle [\![\xi_t]\!]^2\right\rangle = \frac{\partial k_{poly}}{\partial K_t}E_{poly,v}^2 + 4\frac{\partial\mu_{poly}}{\partial K_t}E_{poly,d}^2 \tag{B.29}$$

When considering $\langle T_t^2\rangle = K_t^2\left\langle [\![\xi_t^2]\!]\right\rangle$ according to (B.9), $\partial k_{poly}/\partial K_t = 0$ according to (B.13), and

$\Sigma_{poly,d} = 2\mu_{poly}E_{poly,d}$, (B.29) reduces to

$$\frac{\mathcal{I}}{V_{poly}}\left\langle T_t^2\right\rangle = -\frac{\partial}{\partial K_t}\left(\frac{1}{\mu_{poly}}\right)K_t^2\Sigma_{poly,d}^2 \tag{B.30}$$

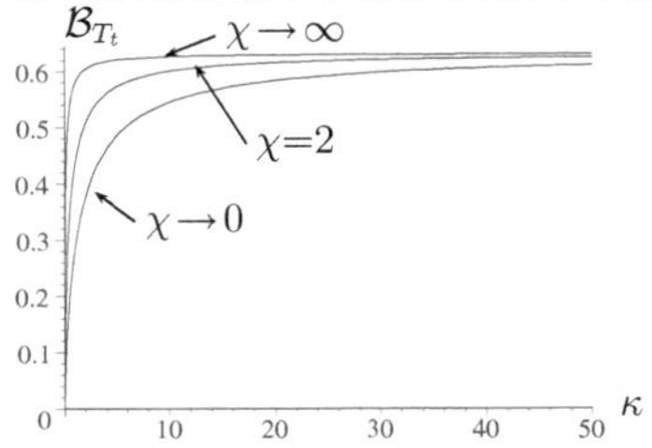

Figure B.5: Concentration factor $\mathcal{B}_{T_t}$ relating 'macroscopic' deviatoric stress on polycrystal to effective tangential traction in intercrystalline interfaces, as function of dimensionless quantity $\kappa = K_t' a/\mu_C$ (interfacial rigidity), for different crystal compressibilities $\chi = \mu_C/k_C$, Eq.(B.32).

where $\Sigma_{poly,d}$ is the equivalent deviatoric stress of the 'macroscopic' second-order stress tensor $\mathbf{\Sigma}_{poly}$,

$$\begin{aligned} \Sigma_{poly,d} &= \sqrt{\frac{1}{2}\mathbf{\Sigma}_{poly,d} : \mathbf{\Sigma}_{poly,d}} \\ \text{with } \mathbf{\Sigma}_{poly,d} &= \mathbf{\Sigma}_{poly} - \Sigma_{poly,m}\mathbf{1}, \\ \text{and } \Sigma_{poly,m} &= \frac{1}{3}\mathrm{tr}\mathbf{\Sigma}_{poly} \end{aligned} \tag{B.31}$$

Combination of (B.30) with $\mathcal{I}/V_{poly} = 3/(2a)$ and with $\kappa = K_t' a/\mu_C$ yields

$$\begin{aligned} &\sqrt{\langle T_t^2 \rangle} = \mathcal{B}_{T_t}\Sigma_{poly,d} \\ &\text{with } \mathcal{B}_{T_t}(\chi{=}\frac{\mu_C}{k_C}, \kappa) = \sqrt{-\frac{1}{3}\kappa^2\frac{\partial}{\partial\kappa}\left(\frac{\mu_C}{\mu_{poly}}\right)} \end{aligned} \tag{B.32}$$

Remarkably, the second-order moment of tangential tractions over all interfaces within the RVE, $\sqrt{\langle T_t^2 \rangle}$, is proportional to the 'macroscopic' equivalent deviatoric stress $\Sigma_{poly,d}$, expressed by the proportionality factor $\mathcal{B}_{T_t}$. The more compressible the solid crystal (the larger $\chi = \mu_C/k_C$), the higher the tangential traction peaks in the intercrystalline interface, generated by an equivalent deviatoric 'macroscopic' stress $\Sigma_{poly,d}$. However, the corresponding concentration factor $\mathcal{B}_{T_t}$ is bounded by $\sqrt{2/5}$ (Fig. B.5),

$$\lim_{\chi\to\infty} \mathcal{B}_{T_t}(\kappa) = \sqrt{\frac{2}{5}} \tag{B.33}$$

On the other hand, for any constant crystal compressibility χ, stiffening the interface (enlarging $\kappa = K_t' a/\mu_C$) also increases the peaks of tangential traction force, i.e. the proportionality factor $\mathcal{B}_{T_t}$, again bounded by $\sqrt{2/5}$ (Fig. B.5),

$$\lim_{\kappa\to\infty} \mathcal{B}_{T_t}(\chi) = \sqrt{\frac{2}{5}} \tag{B.34}$$

Use of the micro traction-macro stress relationships (B.22) and (B.32) in the local interface criterion (B.8) yields a 'macroscopic' polycrystal-specific brittle-failure criterion in the form

$$\mathcal{B}_{T_t}\Sigma_{poly,d} \leq \alpha(h - \Sigma_{poly,m}) \tag{B.35}$$

(B.35) expresses that Coulomb-type brittle failure (B.8) in the interfaces between spherical crystals inside the RVE results in Drucker-Prager-type (brittle) failure properties at the scale of polycrystal.

B.4 Micromechanics of porous material with polycrystalline skeleton

We consider an RVE V_{PORO} [Fig. B.2(b) and Fig. B.9(b)] of a porous material (with porosity φ) where the contiguous solid phase [volume V_S, $V_S = V_{PORO}(1-\varphi)$] is a polycrystal with weak interfaces according to Section B.3. The Mori-Tanaka homogenization scheme has been proven as suitable tool to upscale the elastic properties of the solid phase [k_{poly} and μ_{poly} defined through (B.13), (B.20), (B.21)] to the stiffness of such a porous material, see e.g. (Dormieux 2005; Dormieux et al. 2006b),

$$\mathbb{C}_{PORO} = (1-\varphi)\mathbb{C}_{poly} : \left((1-\varphi)\mathbb{I} + \varphi(\mathbb{I}-\mathbb{S})^{-1}\right)^{-1} \tag{B.36}$$

with the Eshelby tensor $\mathbb{S}$ for spherical inclusions reading as (Eshelby 1957)

$$\mathbb{S} = \frac{3\,k_{poly}}{3k_{poly} + 4\mu_{poly}}\mathbb{J} + \frac{6(k_{poly} + 2\mu_{poly})}{5(3k_{poly} + 4\mu_{poly})}\mathbb{K} \tag{B.37}$$

so that

$$k_{PORO} = \frac{4k_{poly}\mu_{poly}(1-\varphi)}{3k_{poly}\,\varphi + 4\mu_{poly}} \tag{B.38}$$

$$\mu_{PORO} = \mu_{poly}\frac{(1-\varphi)(9k_{poly} + 8\mu_{poly})}{9k_{poly}(1+\frac{2}{3}\varphi) + 8\mu_{poly}(1+\frac{3}{2}\varphi)} \tag{B.39}$$

We consider brittle failure of the overall porous medium if the polycrystal failure criterion (B.35) is reached in highly stressed regions of the polycrystalline matrix. The corresponding ('micro'-)heterogeneity within the solid matrix has recently been shown (Dormieux et al. 2002) to be reasonably considerable through so-called (homogeneous) *effective* ('micro'-) stresses, such as the square root of the spatial average over the solid material phase, of the squares of equivalent deviatoric ('micro'-)stresses,

$$\sqrt{\langle\sigma_d^2\rangle_S} = \sqrt{\frac{1}{V_S}\int_{V_S}\frac{1}{2}\boldsymbol{\sigma}_d(\underline{x}) : \boldsymbol{\sigma}_d(\underline{x})\,\mathrm{d}V} \tag{B.40}$$

$$\text{with} \quad \boldsymbol{\sigma}_d(\underline{x}) = \boldsymbol{\sigma}(\underline{x}) - \frac{1}{3}\mathrm{tr}\,\boldsymbol{\sigma}(\underline{x})\,\mathbf{1} \tag{B.41}$$

The effective deviatoric stress (B.40), used to approximate $\Sigma_{poly,d}$ in (B.35), is accessible through energy considerations similar to those of (B.24) to (B.30), and result to be ((Dormieux et al. 2002), (Dormieux 2005, p. 132))

$$\Sigma^2_{poly,d} \approx \langle \sigma^2_d \rangle_S = \left[-\frac{\partial}{\partial \mu_{poly}} (\frac{1}{k_{PORO}}) \Sigma^2_{PORO,m} - \right.$$
$$\left. -\frac{\partial}{\partial \mu_{poly}} (\frac{1}{\mu_{PORO}}) \Sigma^2_{PORO,d} \right] \frac{\mu^2_{poly}}{(1-\varphi)} \tag{B.42}$$

In analogy to (B.23), the effective mean stress level in the solid matrix is chosen as the stress average over the solid phase,

$$\Sigma_{poly,m} \approx \langle \sigma_m \rangle_S = \frac{1}{V_S} \int_{V_S} \frac{1}{3} \mathrm{tr}\, \boldsymbol{\sigma}(\underline{x})\, \mathrm{d}V =$$
$$= \frac{\Sigma_{PORO,m}}{1-\varphi} \tag{B.43}$$

Use of Eqs.(B.43) and (B.42), together with (B.38) to (B.41), (B.13), and (B.20), in (B.35) yields a failure criterion at the scale of the porous material with polycrystalline interfaces in the solid phase,

$$\left[\frac{3\varphi}{4} - \left(\frac{\alpha}{\mathcal{B}_{T_t}} \right)^2 \right] \Sigma^2_{PORO,m} +$$
$$+ \left[\frac{2\varphi\,(23 - 50\nu_{poly} + 35\nu^2_{poly})}{(-7 + 5\nu_{poly})^2} + 1 \right] \Sigma^2_{PORO,d} +$$
$$+2 \left(\frac{\alpha}{\mathcal{B}_{T_t}} \right)^2 h(1-\varphi)\, \Sigma_{PORO,m} =$$
$$= \left(\frac{\alpha}{\mathcal{B}_{T_t}} \right)^2 h^2 (1-\varphi)^2 \tag{B.44}$$

with

$\nu_{poly} = \nu_{poly}(k_{poly}, \mu_{poly})$ according to (B.15),

$\mu_{poly} = \mu_{poly}(k_C, \mu_C, \kappa)$ according to (B.20),

and $\mathcal{B}_{T_t} = \mathcal{B}_{T_t}(\chi = \frac{\mu_C}{k_C}, \kappa)$ according to (B.32).

The elastic stress domain of the porous medium the matrix of which is a polycrystal with brittle interfaces increases with decreasing crystal compressibility χ (Fig. B.6). For the incompressible

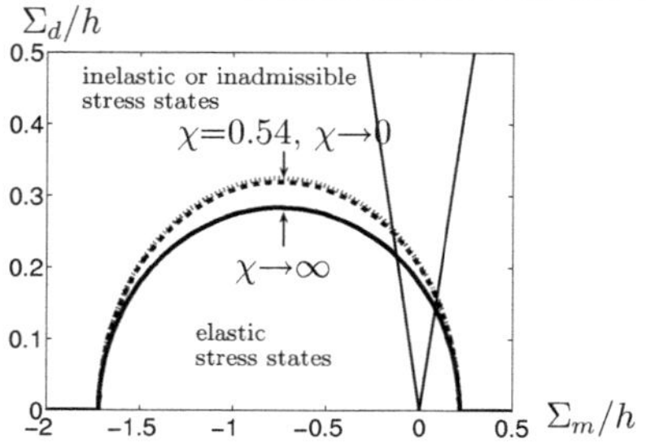

Figure B.6: Elastic limits of a porous material the matrix of which is a polycrystal with brittle interfaces, for different crystal compressibilities $\chi = \mu_C/k_C$ [Eq.(B.44)]: φ=0.5, α=0.3, $\kappa\rightarrow\infty$. Uniaxial load path indicated (thin solid line).

limit case, $\chi\rightarrow 0$, (B.44) reduces to

$$\left[\frac{3\varphi}{4} - \left(\frac{\alpha}{\mathcal{B}_{T_t}}\right)^2\right] \Sigma^2_{PORO,m} + \left(1 + \frac{2}{3}\varphi\right) \Sigma^2_{PORO,d} +$$

$$+2\left(\frac{\alpha}{\mathcal{B}_{T_t}}\right)^2 h(1-\varphi)\,\Sigma_{PORO,m} =$$

$$= \left(\frac{\alpha}{\mathcal{B}_{T_t}}\right)^2 h^2(1-\varphi)^2 \qquad \text{(B.45)}$$

For a crystal compressibility of hydroxyapatite, $\chi\approx 0.54$ (see also Section B.5), the elastic domain increases with decreasing interfacial rigidity (Fig. B.7) and with increasing friction angle α (Fig. B.8). High interfacial rigidities κ or low friction angles α result in closed elastic domains, indicating possible failure of the porous material even under hydrostatic stress states $\mathbf{\Sigma} = \mathbf{1}\Sigma_m$, while low interfacial rigidities κ or high friction angles α are related to open elastic domains, related to infinite resistance of the porous material, as long as the macroscopic stress state $\mathbf{\Sigma}$ contains a certain hydrostatic amount (Figs. B.7 and B.8).

B.5 Application to hydroxyapatite biomaterials

Porous hydroxyapatite (HA) biomaterials are widely used for replacement of hard tissue defects, because of their chemical composition, microstructure and Young's modulus being similar to the bone mineral, called carbonated or calcium-deficient hydroxyapatite (CDHA) (Suchanek and Yoshimura 1998; LeGeros 2002; Hench and Jones 2005). If porous scaffolds are used as bone replacement material in highly loaded anatomical locations, reliability of their mechanical properties is particularly important for the performance of the implants. Therefore, the prediction of strength of HA biomaterials from their microstructure and porosity is of particular interest. To the knowledge of the authors, corresponding micromechanical approaches

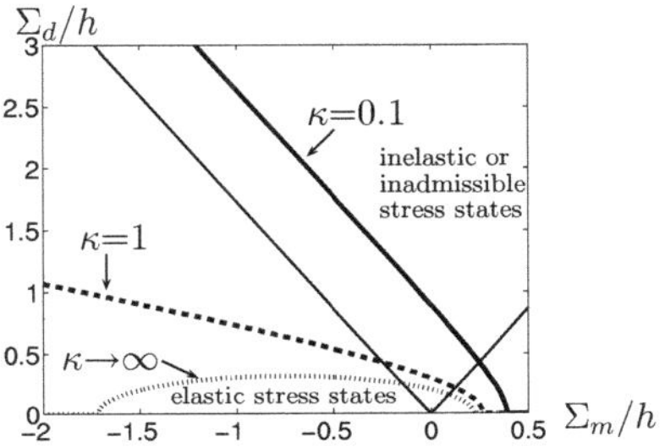

Figure B.7: Elastic limits of a porous material the matrix of which is a polycrystal with brittle interfaces, for different dimensionless quantities $\kappa = K_t' a/\mu_C$ (interfacial rigidity) [Eq.(B.44)]: φ=0.5, α=0.3, χ=0.54. Uniaxial load path indicated (thin solid line).

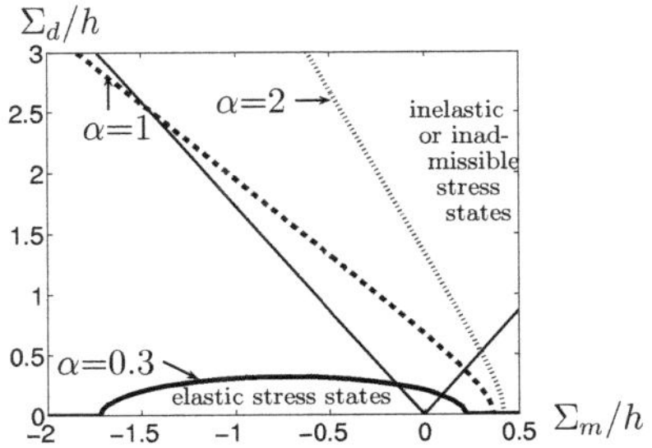

Figure B.8: Elastic limits of a porous material the matrix of which is a polycrystal with brittle interfaces, for different friction angles α [Eq.(B.44)]: φ=0.5, $\kappa \to \infty$, χ=0.54. Uniaxial load path indicated (thin solid line).

(Peelen et al. 1978) φ [%]	f_c [MPa]	(Akao et al. 1981) φ [%]	f_c [MPa]	(Martin and Brown 1995) φ [%]	f_c [MPa]
36	160	2.8	509	27	172.5[a]
48	114	3.9	465	39	119[a]
60	69	9.1	415		
65	45	19.4	308		
70	30				

[a] mean value calculated from three experiments

Table B.2: Experimental data: Compressive strength f_c as function of porosity φ, for artificial hydroxyapatite produced through different synthesis routes.

are extremely rare or inexistent, so that we check in this Section, to which extent the model developed before can serve the purpose of the aforementioned prediction.

B.5.1 Materials processing and uniaxial mechanical testing

We here consider the following artificially produced HA materials:

Peelen et al. (1978) controlled the porosity of HA by a variation of the sintering temperature (1100-1400°C, Table B.2). Compacted commercially available powders were used to produce HA with porosities between 36 and 70%. Cylindrical samples (diameter: 1 cm, length: 1-1.5 cm) were tested in compression (Table B.2).

Akao et al. (1981) precipitated HA powder and sintered it at different temperatures (1150-1300°C). Porosities ranged from 3 to 19% (Table B.2). Compression tests were performed on bars with dimensions of 5x5x10 cm^3 (Table B.2).

Martin and Brown (1995) prepared calcium-deficient HA formed in aqueous solutions at physiological temperatures. The authors realized two different liquid-to-solid weight ratios, resulting in porosities of 27% and 39%, respectively (Table B.2). Cylindrical samples with diameter of ~6 mm were tested in compression (Table B.2).

B.5.2 Micromechanical representation of hydroxyapatite biomaterials

In the hierarchical organization of synthetic hydroxyapatite ceramics, we identify two different scales which will be considered in the framework of a two-step homogenization scheme. The first homogenization step refers to an observation scale of several hundreds of microns where hydroxyapatite crystals are separated by boundaries or interfaces [Fig. B.9(a)]. The latter will be shown to be a potential nucleus for failure of the material. The corresponding homogenized material is called 'hydroxyapatite polycrystal with interfaces'. At the microstructural scale with a characteristic length of some millimeters [Fig. B.9(b)], pores are embedded in a matrix which is made up of the material which was homogenized in the first upscaling step.

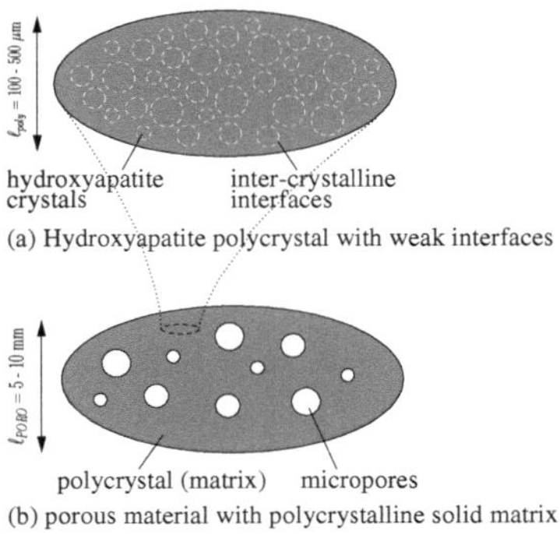

Figure B.9: Micromechanical representation of a porous hydroxyapatite polycrystal by means of a two-step homogenization procedure.

B.5.3 Elastic properties of single crystals of hydroxyapatite

An ultrasonic interferometer technique delivers typical values for bulk and shear moduli, $k_C = k_{HA} = 82.6$ GPa and $\mu_C = \mu_{HA} = 44.9$ GPa (Katz and Ukraincik 1971).

B.5.4 Biomaterial-independent properties of interfaces between hydroxyapatite crystals, α, h, κ – back-analysis

The expression for macroscopic admissible stress states (B.44) contains three material properties which are difficult to be directly accessed, namely the friction angle α, the cohesion h, and the rigidity κ of the interfaces. Therefore, these phase properties will be determined by means of an optimization procedure providing the closest match of model predictions to experimentally determined uniaxial compressive strength data of hydroxyapatite biomaterials, given in Table B.2 (Peelen et al. 1978; Akao et al. 1981; Martin and Brown 1995).

The sum of squares of relative errors between predicted strength and experimental strength values is minimized,

$$\mathcal{G}(\alpha, h, \kappa) = \sum_i \left(\frac{f_{c,i}^{pred} - f_{c,i}^{exp}}{f_{c,i}^{exp}} \right)^2 \rightarrow 0 \qquad \text{(B.46)}$$
$$\Rightarrow \alpha^{opt}, h^{opt}, \kappa^{opt}$$

where $f_{c,i}^{pred}$ denotes predicted uniaxial compressive strength values obtained from Eq.(B.44) with $\Sigma_{PORO,m} = -f_{c,i}^{pred}/3$, $\Sigma_{PORO,d} = f_{c,i}^{pred}/\sqrt{3}$, together with Eqs.(B.13), (B.20), and (B.32), for porosity values φ_i according to Table B.2. $f_{c,i}^{exp}$ is the corresponding i-th experimental strength value, see Table B.2.

We use the 'two-membered evolution strategy' (Schwefel 1977; Hellmich and Ulm 2002b), closely related to the ideas of Darwin's evolution theory. The components of a three-dimensional vector of estimations for α, h and κ, $(\alpha, h, \kappa)_{parent}$, representing the 'parent', are slightly varied by help of a random number generator (representing 'mutations'), resulting in a vector $(\alpha, h, \kappa)_{child}$, representing the 'child',

$$(\alpha, h, \kappa)_{child} = (\alpha, h, \kappa)_{parent} + \\ +(\mathcal{N}\sigma\alpha_{parent}, \mathcal{N}\sigma h_{parent}, \mathcal{N}\sigma\kappa_{parent}) \tag{B.47}$$

$\mathcal{N}$ denotes a number produced by a standardized normally distributed random number generator standardly available in MATLAB (Hunt et al. 2001). σ stands for a scattering factor which will be dealt with later on.

If the child fits better in its 'environment' than the parent, i.e., if

$$\mathcal{G}[(\alpha, h, \kappa)_{child}] < \mathcal{G}[(\alpha, h, \kappa)_{parent}] \tag{B.48}$$

see (B.46), vector $(\alpha, h, \kappa)_{child}$ will be further varied, i.e., it then becomes the parent for the next generation. If not, the original parent undergoes new mutations.

Based on the number of 'successes' of the evolution, i.e., the number of cases for which Eq.(B.48) holds, the scattering factor σ is changed: If the total number of successes within the last 10 mutations exceeds a certain threshold (typically 4), σ is enlarged, otherwise it is reduced.

If the difference between $\mathcal{G}[(\alpha, h, \kappa)_{parent}]$ and $\mathcal{G}[(\alpha, h, \kappa)_{child}]$ lies within a prescribed tolerance over a certain number of mutations, the optimum $(\alpha^{opt}, h^{opt}, \kappa^{opt}) \approx (\alpha, h, \kappa)_{parent} \approx (\alpha, h, \kappa)_{child}$ has been reached.

Applying this procedure, (B.46) to (B.48), to (B.44) and using the experimental data from Table B.2 yields, depending on the start values of the optimization procedure, a set of solution vectors $(\alpha^{opt}, h^{opt}, \kappa^{opt})$ which are equal in terms of the highly satisfactory correlation coefficient $(r^2 = 0.97)$ between the respective model predictions and the corresponding experimental data for uniaxial compressive strength (see Fig. B.10). To give an example, $(\alpha^{opt}, h^{opt}, \kappa^{opt})=$ (0.6750, 17.2397, 0.9119) and (0.9345, 17.7664, 6.4160) (h has the dimension [MPa]) are two of these solution vectors. For all calculated 'optimal' solution vectors, we find a constant ratio $\alpha' = \alpha/\mathcal{B}_{T_t} = 1.61$ [see Eqs. (B.32) and (B.20)], implying a relationship between α and κ, depicted in Fig. B.11.

Clearly, it would be interesting to cross-check these interface failure parameters derived from our 'inverse method' with other direct tests. Deplorably, an extensive literature check could not provide any direct in situ measurements of stresses and failure mechanisms at the interface 'micro' level. The only additional experimental evidence are scanning electron micrographs (Fig. 2 in Ref. (De With et al. 1981), Figs. 5-7 in Ref. (Martin and Brown 1995)): These images, however, clearly show sharp, rough failure surfaces, coinciding with the boundaries of single, micrometer-sized grains. This, together with the sharp stress drops in corresponding ('macroscopic') stress-strain diagrams indicating brittle overall failure, strongly suggests brittle failure of the crystal interfaces, as we have modelled herein.

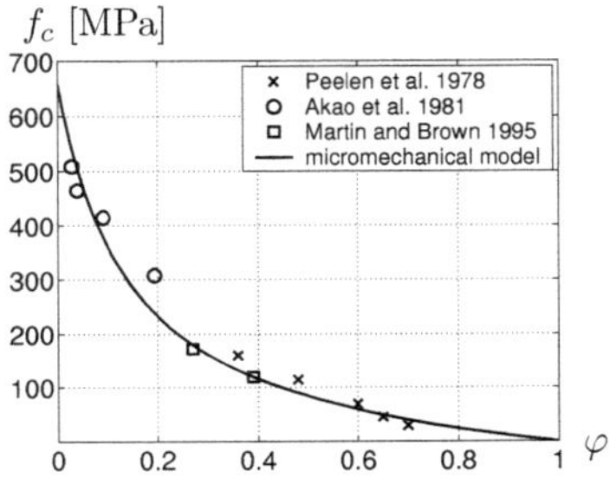

Figure B.10: Uniaxial compressive strength f_c of porous hydroxyapatite biomaterial as function of porosity φ: Model prediction according to Eq.(B.44) or Eq.(B.49), evaluated with $\Sigma_{PORO,m}=-f_c/3$, $\Sigma_{PORO,d}=f_c/\sqrt{3}$, compared to experimental data (Table B.2). Correlation coefficient $r^2=0.97$.

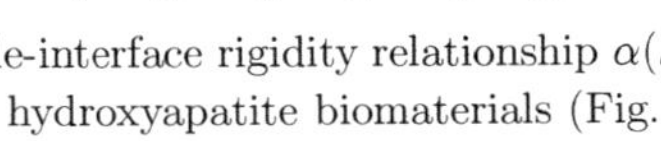

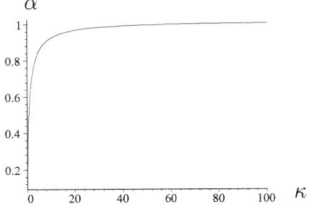

Figure B.11: Friction angle-interface rigidity relationship $\alpha(\kappa)$ suitable for representation of strength of hydroxyapatite biomaterials (Fig. B.10).

B.5.5 Brittle versus ductile failure of solid matrix in porous medium

From a purely mathematical standpoint, it is interesting to compare the elastic domain (B.44) to the yield surface of a porous medium, related to failure of a ductile (not a brittle) solid matrix obeying Drucker-Prager criterion (B.35). This yield surface can be obtained through non-linear homogenization based on effective quantities (B.42) and (B.43), as detailed in (Dormieux 2005; Dormieux et al. 2006b),

$$
\begin{aligned}
&(\frac{3\varphi}{4} - \alpha'^2)\Sigma^2_{PORO,m} + (1 + \frac{2}{3}\varphi)\Sigma^2_{PORO,d} + \\
&+2\alpha'^2 h(1-\varphi)\Sigma_{PORO,m} = \alpha'^2 h^2 (1-\varphi)^2
\end{aligned}
\tag{B.49}
$$

with $\alpha' = \alpha/\mathcal{B}_{T_t}(\kappa)$ and h as only two parameters being left for an optimization procedure to match the experimental data of Fig. B.10 and Table B.2. This procedure delivers a cohesion $h^{opt} = 16.51$ MPa (close to the values obtained for the brittle case in Section B.5.4) and ratio $\alpha'^{opt} = 1.61$ which is quasi-identical to the one obtained for the brittle case (Section B.5.4), implying an α-κ-relationship quasi-identical to that of Fig. B.11. This means that the failure of porous hydroxyapatite biomaterials can be equally well represented by a brittle elastic-limit-type micromechanics model and a ductile one related to limit analysis. In this context, it is very interesting to note that the *ductile* criterion (B.49) is even *identical* to the elastic domain for *incompressible* solid matrices, Eq.(B.45).

Accordingly, one might argue that the nature of the heterogeneity of the stresses in the solid matrix (considered herein by quadratic averages) is far more important for the overall failure of the material than the precise mode of local interface failure (brittle or ductile). However, as regards hydroxyapatite biomaterials, experiments (Chu et al. 2002; Martin and Brown 1995; Pramanik et al. 2007) strongly support brittle failure: A comprehensive mechanical formulation for its possible origin, namely breaking of weak interfaces between hydroxyapatite crystals, was the main focus of the present paper.

B.6 Appendix: solution of matrix-inclusion problem with compliant interface ('generalized Eshelby problem', Fig. B.3)

Solution of Eqs.(B.12), (B.13), (B.14), and (B.16) for the constants B_{ex}, C_{ex}, A_{in}, and B_{in} yields them as:

$$B_{ex} = -a^5(176\mu_{poly}^3\mu_C^2 + 24\mu_{poly}^2\mu_C^3$$
$$-12\mu_C^3 a\, k_{poly}K_t - 171\mu_C^2 a\, k_{poly}K_t k_C + 240\mu_{poly}^3 k_C\mu_C$$
$$+136\mu_{poly}^3\mu_C aK_t + 48\mu_{poly}^3 aK_t k_C - 132a\mu_{poly}^2 K_t\mu_C^2$$
$$+528\mu_{poly}^2 k_{poly}\mu_C^2 + 9a\mu_{poly}^2 K_t k_C\mu_C$$
$$+720\mu_{poly}^2 k_{poly}k_C\mu_C + 342\mu_{poly}^2\mu_C^2 k_C$$
$$+144\mu_{poly}^2 ak_{poly}K_t k_C + 408\mu_{poly}^2 ak_{poly}K_t\mu_C$$
$$+27aK_t k_C k_{poly}\mu_{poly}\mu_C - 396aK_t\mu_C^2 k_{poly}\mu_{poly}$$
$$-57\mu_{poly}\mu_C^2 aK_t k_C - 4\mu_{poly}\mu_C^3 aK_t)/\mathcal{N} \tag{B.50}$$

$$C_{ex} = 5a^3(48\mu_{poly}^2 aK_t k_C + 240\mu_{poly}^2 k_C\mu_C$$
$$+136\mu_{poly}^2 aK_t\mu_C + 176\mu_{poly}^2\mu_C^2 - 8\mu_{poly}\mu_C^3$$
$$+9\mu_{poly}aK_t k_C\mu_C - 114\mu_{poly}\mu_C^2 k_C - 132\mu_{poly}aK_t\mu_C^2$$
$$-57\mu_C^2 aK_t k_C - 4\mu_C^3 aK_t)\mu_{poly}/\mathcal{N} \tag{B.51}$$

$$A_{in} = 5(544\mu_{poly}^2 aK_t\mu_C + 192\mu_{poly}^2 aK_t k_C + 320\mu_{poly}^2\mu_C^2$$
$$+1536\mu_{poly}^2 k_C\mu_C + 16\mu_{poly}aK_t\mu_C^2 + 228\mu_{poly}aK_t k_C\mu_C$$
$$+408a\, k_{poly}\mu_{poly}K_t\mu_C + 144a\, k_{poly}\mu_{poly}K_t k_C$$
$$+240k_{poly}\mu_{poly}\mu_C^2 + 1152k_{poly}\mu_{poly}k_C\mu_C$$
$$+12aK_t\mu_C^2 k_{poly} + 171aK_t k_C k_{poly}\mu_C)\mu_{poly}/\mathcal{N} \tag{B.52}$$

$$B_{in} = 240\mu_{poly}^2\mu_C(8\mu_{poly}\mu_C + 6k_{poly}\mu_C$$
$$-12\mu_{poly}k_C - 9k_{poly}k_C)/(a^2\mathcal{N}) \tag{B.53}$$

$$\mathcal{N} = 1408\mu_{poly}^3\mu_C^2 + 192\mu_{poly}^2\mu_C^3 + 24\mu_C^3 a k_{poly} K_t$$
$$+342\mu_C^2 a\, k_{poly} K_t k_C + 1920\mu_{poly}^3 k_C \mu_C$$
$$+1088\mu_{poly}^3\mu_C a K_t + 384\mu_{poly}^3 a K_t k_C$$
$$+1664 a\mu_{poly}^2 K_t \mu_C^2 + 1584\mu_{poly}^2 k_{poly}\mu_C^2$$
$$+1032 a\mu_{poly}^2 K_t k_C \mu_C + 2160\mu_{poly}^2 k_{poly} k_C \mu_C$$
$$+2736\mu_{poly}^2\mu_C^2 k_C + 432\mu_{poly}^2 a\, k_{poly} K_t k_C$$
$$+1224\mu_{poly}^2 a\, k_{poly} K_t \mu_C + 801 a K_t k_C k_{poly}\mu_{poly}\mu_C$$
$$+852 a K_t \mu_C^2 k_{poly}\mu_{poly} + 1710\mu_{poly}\mu_C^2 k_{poly} k_C$$
$$+120\mu_{poly}\mu_C^3 k_{poly} + 684\mu_{poly}\mu_C^2 a K_t k_C$$
$$+48\mu_{poly}\mu_C^3 a K_t \qquad \text{(B.54)}$$

They define the displacement fields (B.14) and (B.16), which give access to strains $\boldsymbol{\varepsilon} = \nabla^s \underline{\xi}$, stresses $\boldsymbol{\sigma}$ (via $(B.12)_1$, and $(B.12)_3$ respectively), mean interface displacements $\bar{\underline{\xi}}$ and interface tractions $\underline{T}$ (via $(B.12)_2$).

Publication C

Mechanical behavior of hydroxyapatite biomaterials: An experimentally validated micromechanical model for elasticity and strength (Fritsch et al. 2009a)

Authored by Andreas Fritsch, Luc Dormieux, Christian Hellmich, and Julien Sanahuja
Published in *Journal of Biomedical Materials Research Part A*, Volume 88A, pages 149–161

Hydroxyapatite biomaterials production has been a major field in biomaterials science and biomechanical engineering. As concerns prediction of their stiffness and strength, we propose to go beyond statistical correlations with porosity or empirical structure-property relationships, as to resolve the material-immanent microstructures governing the overall mechanical behavior. The macroscopic mechanical properties are estimated from the microstructures of the materials and their composition, in a homogenization process based on continuum micromechanics. Thereby, biomaterials are envisioned as porous polycrystals consisting of hydroxyapatite needles and spherical pores. Validation of respective micromechanical models relies on two independent experimental sets: Biomaterial-specific macroscopic (homogenized) stiffness and uniaxial (tensile and compressive) strength predicted from biomaterial-specific porosities, on the basis of biomaterial-independent ('universal') elastic and strength properties of hydroxyapatite, are compared to corresponding biomaterial-specific experimentally determined (acoustic and mechanical) stiffness and strength values. The good agreement between model predictions and the corresponding experiments underlines the potential of micromechanical modeling in improving biomaterial design, through optimization of key parameters such as porosities or geometries of microstructures, in order to reach desired values for biomaterial stiffness or strength.

C.1 Introduction

Hydroxyapatite [HA, with chemical formula $Ca_{10}(PO_4)_6(OH)_2$ in its pure ('stoichiometric') form] biomaterials production has been a major field in biomaterials science and biomechanical engineering due to their excellent biocompatibility, and since their chemical composition, structure, and mechanical properties are similar to bone mineral (Hench and Jones 2005). Aiming at mimicking the bone mineral and its important biological and mechanical properties within bone tissues, HA is widely used for biomedical applications: They encompass coating of orthopedic and dental implants (Dorozhkin and Epple 2002), artificial hard tissue replacement implants in orthopedics, maxillofacial and dental implant surgery (Charrière et al. 2001). Thereby, HA is used either in a pure state (Frame et al. 1981), (Mastrogiacomo et al. 2006) or as composite, with ceramic, metallic or polymer inclusions as reinforcing component (Verma et al. 2006).

Typical examples for powder-based production of porous hydroxyapatite biomaterials were produced by the following researchers (see also Table C.1):

- Peelen et al. (1978) mixed commercially available HA powders with a 10% hydrogen peroxide solution, poured it into a mold, and controlled the porosity of HA ceramics by a variation of the sintering temperature (Tables C.1 and C.4).
- Akao et al. (1981) precipitated HA powder, which was compacted and sintered at different temperatures (Tables C.1, C.3, and C.4).
- De With et al. (1981) compacted and sintered isostatically pressed HA powder (Tables C.1 and C.3).
- Shareef et al. (1993) produced mixtures with different weight ratios of commercially available fine and coarse powders. Ring-shaped samples were formed by uniaxial pressing and then sintered. (Tables C.1 and C.4).
- Arita et al. (1995) used mixing of starting powders (see Table C.1) and a casting process to produce green bodies made of HA before sintering (Tables C.1 and C.3).
- Martin and Brown (1995) prepared calcium-deficient HA formed in aqueous solutions at physiological temperature. The authors realized two different liquid-to-solid weight ratios, resulting in two different porosities (Tables C.1 and C.4).
- Liu (1998) prepared HA powder by mixing of starting powders (see Table C.1). Water and polyvinyl butyral powder were added to HA before casting the slurry and sintering the green bodies (Tables C.1, C.3, and C.4).
- Charrière et al. (2001) mixed commercially available powders in an aqueous solution and used a casting process to obtain HA cement (Tables C.1 and 3).

The mechanical and microstructural properties, i.e. stiffness/strength and porosity, of these materials (see Tables C.3 and C.4) will be used as to validate the theoretical developments described in this article. Thereby, we will go beyond statistical correlations between porosity and stiffness/strength or empirical structure-property relationships (Rao and Boehm 1974; Driessen et al. 1982; Katz and Harper 1990), as to resolve the material-immanent microstructures governing the overall mechanical behavior, in the theoretical framework of continuum micromechanics.

Literature reference	Source material(s)	Processing steps	Shape/size of samples	Typical pore size	Mechanical characterization method
(Peelen et al. 1978)	Commercially available HA powder	Mixing of HA powder with 10% hydrogen peroxide solution, poured into mold, sintering	Cylindrical (d=1 cm, h=1-1.5 cm)	1-200 μm	Uniaxial, quasi-static compressive test (Table C.4)
(Akao et al. 1981)	$Ca(OH)_2$, H_3PO_4	Mixing of starting powders to precipitate HA powder, mixed with water and cornstarch, compaction, sintering	Bars (5x5x10 cm^3)	~1 μm (pore size ≈ grain size, see also Figs. 2-4 of the reference)	Uniaxial, quasi-static compressive test (Tables C.3 and C.4)
(De With et al. 1981)	Commercially available HA powder	Mixing of HA powder with water, compaction, sintering	Cylindrical (d=5 mm, h=15 mm)	1-5 μm (see Figs. 2 and 7 of the reference)	Ultrasonic pulse-echo technique (Table C.3)
(Shareef et al. 1993)	Commercially available fine and coarse HA powders	Mixing of HA powders, compaction, sintering	Ring-shaped (inner diameter 34 mm)	1 μm	Quasi-static tensile test (Stanford ring bursting test, Table C.4)
(Arita et al. 1995)	$CaHPO_4$, $CaCO_3$	Mixing of starting powders with water, tape casting, sintering	Discs (d=2.54 cm)	~1 μm (see Fig. 6 of the reference)	Resonance frequency method (Table C.3)
(Martin and Brown 1995)	$CaHPO_4$, $Ca_4(PO_4)_2O$	Mixing of starting powders with water, precipitation, compaction at low temperature	Cylindrical (d~6.40 mm, h=5.09-6.39 mm)	~1-2 μm	Uniaxial, quasi-static compressive test (Table C.4)
(Liu 1998)	$Ca(OH)_2$, H_3PO_4	Mixing of starting powders in solution, mixing of HA powder with water and polyvinyl butyral powder in a slurry, slip casting, sintering	Bars (5x8x50 mm^3)	2-200 μm	Quasi-static tensile test (three-point bending; Tables C.3 and C.4)
(Charrière et al. 2001)	$CaHPO_4$, $CaCO_3$	Mixing of starting powders with polyacrylic acid solution in suspension, poured into mold, slip casting	Hollow cylinders (d=18 mm, h=40 mm)	~1 μm	Uniaxial, quasi-static compressive test (Table C.3)

Table C.1: Hydroxyapatite-based porous biomaterials used for model validation: survey on processing, pore size, and mechanical characterization methods.

C.2 Fundamentals of continuum micromechanics

C.2.1 Representative volume element and phase properties

In continuum micromechanics (Hill 1963; Hashin 1983; Suquet 1997a; Zaoui 2002), a material is understood as a macrohomogeneous, but microheterogeneous body filling a representative volume element (RVE) with characteristic length ℓ, $\ell \gg d$, d standing for the characteristic length of inhomogeneities within the RVE, and $\ell \ll \mathcal{L}$, $\mathcal{L}$ standing for the characteristic lengths of geometry or loading of a structure built up by the material defined on the RVE (Fig. C.1). In general, the microstructure within one RVE is so complicated that it cannot be described in complete detail. Therefore, N_r quasi-homogeneous subdomains with known physical quantities are reasonably chosen. They are called material phases [Fig. C.1(a)].

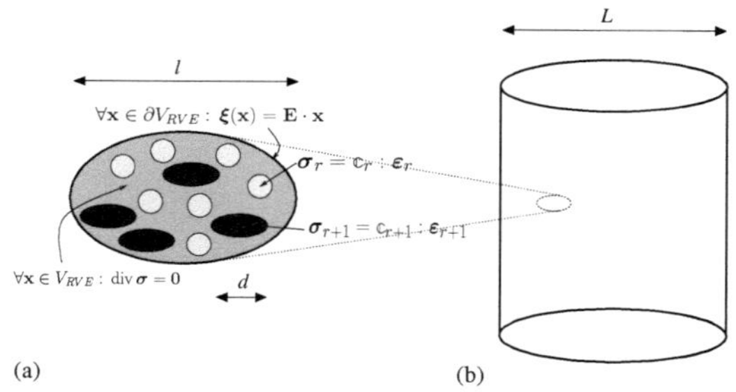

Figure C.1: (a) Loading of a representative volume element, built up by phases r with stiffness $\mathbb{C}_r$ and strength properties $\mathfrak{f}(\boldsymbol{\sigma}) = 0$, according to continuum micromechanics (Hashin 1983; Zaoui 2002): Displacements $\boldsymbol{\xi}$, related to a constant (homogenized) strain $\boldsymbol{E}$, are imposed at the boundary of the RVE; (b) structure built up of material defined on RVE (a).

Quantitative phase properties are volume fractions f_r of phases $r = 1, \ldots, N_r$, as well as elastic and strength properties of phases. As regards phase elasticity, the fourth-order stiffness tensor $\mathbb{C}_r$ relates the (average microscopic) second-order strain tensor in phase r, $\boldsymbol{\varepsilon}_r$, to the (average microscopic) second-order stress tensor in phase r, $\boldsymbol{\sigma}_r$,

$$\boldsymbol{\sigma}_r = \mathbb{C}_r : \boldsymbol{\varepsilon}_r \tag{C.1}$$

As regards phase strength, brittle failure can be associated to the boundary of an elastic domain $\mathfrak{f}_r(\boldsymbol{\sigma}) < 0$,

$$\mathfrak{f}_r(\boldsymbol{\sigma}) = 0 \tag{C.2}$$

defined in the space of microstresses $\boldsymbol{\sigma}(\boldsymbol{x})$, $\boldsymbol{x}$ being the position vector for locations within or at the boundary of the RVE.

Also the spatial arrangement of the phases needs to be specified. In this respect, two cases are of particular interest: (i) one or several inclusion phases with different shapes are embedded in a contiguous 'matrix' phase (as in a reinforced composite material), or (ii) mutual contact of all disorderly arranged phases (as in a polycrystal).

C.2.2 Averaging – Homogenization

The central goal of continuum micromechanics is to estimate the mechanical properties (such as elasticity or strength) of the material defined on the RVE (the macrohomogeneous, but microheterogeneous medium) from the aforementioned phase properties. This procedure is referred to as homogenization or one homogenization step. Therefore, homogeneous (macroscopic) strains $\boldsymbol{E}$ are imposed onto the RVE, in terms of displacements at its boundary ∂V:

$$\forall \boldsymbol{x} \in \partial V : \boldsymbol{\xi}(\boldsymbol{x}) = \boldsymbol{E} \cdot \boldsymbol{x} \tag{C.3}$$

As a consequence, the resulting kinematically compatible microstrains $\boldsymbol{\varepsilon}(\boldsymbol{x})$ throughout the RVE with volume V_{RVE} fulfill the average condition (Hashin 1983),

$$\boldsymbol{E} = \langle \boldsymbol{\varepsilon} \rangle = \frac{1}{V_{RVE}} \int\limits_{V_{RVE}} \boldsymbol{\varepsilon} \, \mathrm{d}V = \sum_r f_r \boldsymbol{\varepsilon}_r \tag{C.4}$$

providing a link between micro and macro strains. Analogously, homogenized (macroscopic) stresses $\boldsymbol{\Sigma}$ are defined as the spatial average over the RVE, of the microstresses $\boldsymbol{\sigma}(\boldsymbol{x})$,

$$\boldsymbol{\Sigma} = \langle \boldsymbol{\sigma} \rangle = \frac{1}{V_{RVE}} \int\limits_{V_{RVE}} \boldsymbol{\sigma} \, \mathrm{d}V = \sum_r f_r \boldsymbol{\sigma}_r \tag{C.5}$$

Homogenized (macroscopic) stresses and strains, $\boldsymbol{\Sigma}$ and $\boldsymbol{E}$, are related by the homogenized (macroscopic) stiffness tensor $\mathbb{C}$,

$$\boldsymbol{\Sigma} = \mathbb{C} : \boldsymbol{E} \tag{C.6}$$

which needs to be linked to the stiffnesses $\mathbb{c}_r$, the shape, and the spatial arrangement of the phases (Section C.2.1). This link is based on the linear relation between the homogenized (macroscopic) strain $\boldsymbol{E}$ and the average (microscopic) strain $\boldsymbol{\varepsilon}_r$, resulting from the superposition principle valid for linear elasticity, see Eq. (C.1) (Hill 1963). This relation is expressed in terms of the fourth-order concentration tensors $\mathbb{A}_r$ of each of the phases r

$$\boldsymbol{\varepsilon}_r = \mathbb{A}_r : \boldsymbol{E} \tag{C.7}$$

Insertion of Eq. (C.7) into Eq. (C.1) and averaging over all phases according to Eq. (C.5) leads to

$$\boldsymbol{\Sigma} = \sum_r f_r \mathbb{c}_r : \mathbb{A}_r : \boldsymbol{E} \tag{C.8}$$

From Eq. (C.8) and Eq. (C.6) we can identify the sought relation between the phase stiffness tensors $\mathbb{c}_r$ and the overall homogenized stiffness $\mathbb{C}$ of the RVE,

$$\mathbb{C} = \sum_r f_r \mathbb{c}_r : \mathbb{A}_r \tag{C.9}$$

The concentration tensors $\mathbb{A}_r$ can be suitably estimated from Eshelby's 1957 matrix-inclusion problem (Eshelby 1957), according to (Zaoui 2002), (Benveniste 1987)

$$\mathbb{A}_r^{est} = \left[\mathbb{I} + \mathbb{P}_r^0 : (\mathbb{c}_r - \mathbb{C}^0)\right]^{-1} : \left\{\sum_s f_s \left[\mathbb{I} + \mathbb{P}_s^0 : (\mathbb{c}_s - \mathbb{C}^0)\right]^{-1}\right\}^{-1} \tag{C.10}$$

where $\mathbb{I}$, $I_{ijkl} = 1/2(\delta_{ik}\delta_{jl} + \delta_{il}\delta_{kj})$, is the fourth-order unity tensor, δ_{ij} (Kronecker delta) are the components of second-order identity tensor $\mathbf{1}$, and the fourth-order Hill tensor $\mathbb{P}_r^0$ accounts for the shape of phase r, represented as an ellipsoidal inclusion embedded in a fictitious matrix of stiffness $\mathbb{C}^0$. For isotropic matrices (which is the case considered throughout this article), $\mathbb{P}_r^0$ is accessible via the Eshelby tensor (Eshelby 1957)

$$\mathbb{S}_r^{esh,0} = \mathbb{P}_r^0 : \mathbb{C}^0 \tag{C.11}$$

see also Section C.3.

Backsubstitution of Eq. (C.10) into Eq. (C.9) delivers the sought estimate for the homogenized (macroscopic) stiffness tensor, $\mathbb{C}^{est}$, as

$$\mathbb{C}^{est} = \sum_r f_r \mathbb{c}_r : \left[\mathbb{I} + \mathbb{P}_r^0 : (\mathbb{c}_r - \mathbb{C}^0)\right]^{-1} : \left\{\sum_s f_s \left[\mathbb{I} + \mathbb{P}_s^0 : (\mathbb{c}_s - \mathbb{C}^0)\right]^{-1}\right\}^{-1} \tag{C.12}$$

Choice of matrix stiffness $\mathbb{C}^0$ determines which type of interactions between the phases is considered: For $\mathbb{C}^0$ coinciding with one of the phase stiffnesses (Mori-Tanaka scheme (Mori and Tanaka 1973)), a composite material is represented (contiguous matrix with inclusions); for $\mathbb{C}^0 = \mathbb{C}^{est}$ (self-consistent scheme (Hershey 1954; Kröner 1958), a dispersed arrangement of the phases is considered (typical for polycrystals).

As long as the average phase strains $\boldsymbol{\varepsilon}_r$ are relevant for brittle phase failure, resulting in overall failure of the RVE, concentration relation (C.7) allows for translation of the brittle failure criterion of the weakest phase $r = w$ into a macroscopic (homogenized) brittle failure criterion, according to (C.1), (C.2), (C.6) and (C.7),

$$\mathfrak{f}_w(\boldsymbol{\sigma}) = 0 = \mathfrak{f}_w(\mathbb{c}_w : \boldsymbol{\varepsilon}_w) = \mathfrak{f}_w(\mathbb{c}_w : \mathbb{A}_w : \boldsymbol{E}) = \mathfrak{f}_w(\mathbb{c}_w : \mathbb{A}_w : \mathbb{C}^{-1} : \boldsymbol{\Sigma}) = \mathfrak{F}(\boldsymbol{\Sigma}) \tag{C.13}$$

Fourth-order tensor operations such as the ones occurring in Eqs. (C.1) and (C.6)-(C.12) can be suitably evaluated in a vector/matrix-based software, through a compressed vector/matrix notation with normalized tensorial basis, often referred to as the Kelvin or the Mandel notation, see e.g. (Cowin and Mehrabadi 1992; Cowin 2003) for details.

C.3 Micromechanical representation of porous biomaterials made of hydroxyapatite – stiffness and strength estimates

In the line of the concept presented in Section C.2, we envision biomaterials made of hydroxyapatite as porous polycrystals consisting of hydroxyapatite needles (Fig. 7 of (Shareef et al. 1993); Fig. 2 of (Liu 1998)) with stiffness $\mathbb{c}_{HA}$ and volume fraction $(1-\phi)$, being oriented in all space directions, and of spherical (empty) pores with vanishing stiffness and volume fraction ϕ (porosity) (see Figs. C.2 and C.3).

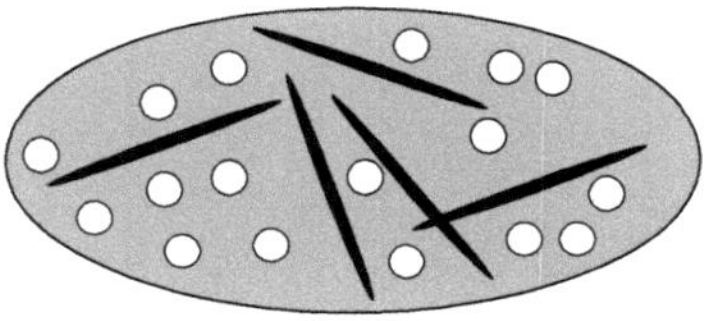

Figure C.2: RVE of polycrystal representing a porous biomaterial made of hydroxyapatite: Uniform orientation distribution of cylindrical (needle-like) inclusions and spherical (empty) pores, in fictitious matrix with stiffness of overall porous polycrystal and vanishing volume fraction.

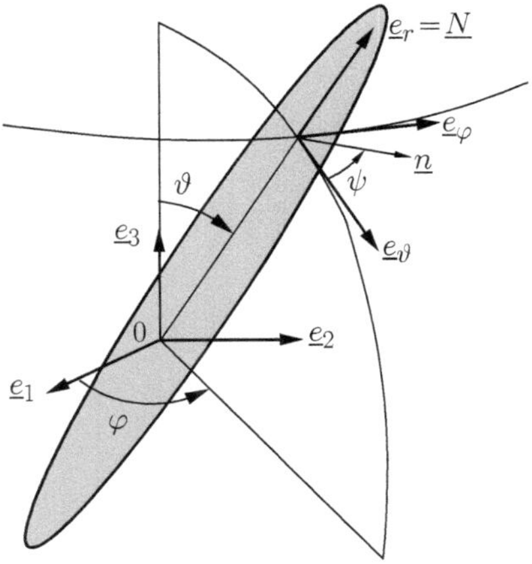

Figure C.3: Cylindrical (needle-like) HA inclusions oriented along vector $\boldsymbol{N}$ and inclined by angles ϑ and φ with respect to the reference frame ($\boldsymbol{e}_1$, $\boldsymbol{e}_2$, $\boldsymbol{e}_3$); local base frame ($\boldsymbol{e}_r$, $\boldsymbol{e}_\varphi$, $\boldsymbol{e}_\vartheta$) is attached to the needle.

C.3.1 Stiffness estimate

In a reference frame ($\boldsymbol{e}_1$, $\boldsymbol{e}_2$, $\boldsymbol{e}_3$), the HA needle orientation vector $\boldsymbol{N} = \boldsymbol{e}_r$ is given by Euler angles ϑ and φ (see Fig. C.3). Specification of Eq. (C.12) for $\mathbb{C}^0 = \mathbb{C}^{est} = \mathbb{C}_{poly}$ (self-consistent scheme) and for an infinite number of solid phases related to orientation directions $\boldsymbol{N} = \boldsymbol{e}_r(\vartheta, \varphi)$, which are uniformly distributed in space ($\varphi \in [0, 2\pi]; \vartheta \in [0, \pi]$), yields the homogenized stiffness of the porous hydroxyapatite biomaterial depicted in Fig. C.2 (Fritsch et al. 2006)

$$\begin{aligned}\mathbb{C}_{poly} &= (1-\phi)\, \mathbb{c}_{HA} : \left\langle \left[\mathbb{I} + \mathbb{P}^{poly}_{cyl} : (\mathbb{c}_{HA} - \mathbb{C}_{poly})\right]^{-1} \right\rangle : \\ &\left\{ (1-\phi) \left\langle \left[\mathbb{I} + \mathbb{P}^{poly}_{cyl} : (\mathbb{c}_{HA} - \mathbb{C}_{poly})\right]^{-1} \right\rangle + \phi(\mathbb{I} - \mathbb{P}^{poly}_{sph} : \mathbb{C}_{poly})^{-1} \right\}^{-1} \end{aligned} \quad \text{(C.14)}$$

with the angular average

$$\begin{aligned}&\left\langle [\mathbb{I} + \mathbb{P}^{poly}_{cyl} : (\mathbb{c}_{HA} - \mathbb{C}_{poly})]^{-1} \right\rangle = \\ &= \int_{\varphi=0}^{2\pi} \int_{\vartheta=0}^{\pi} \left[\mathbb{I} + \mathbb{P}^{poly}_{cyl}(\vartheta, \varphi) : (\mathbb{c}_{HA} - \mathbb{C}_{poly})\right]^{-1} \frac{\sin\vartheta \, \mathrm{d}\vartheta \, \mathrm{d}\varphi}{4\pi} \end{aligned} \quad \text{(C.15)}$$

$\mathbb{P}^{poly}_{sph}$ and $\mathbb{P}^{poly}_{cyl}$ are the fourth-order Hill tensors for spherical and cylindrical inclusions, respectively, in an isotropic matrix with stiffness $\mathbb{C}_{poly} = 3k_{poly}\mathbb{J} + 2\mu_{poly}\mathbb{K}$; $\mathbb{J}$, $J_{ijkl} = 1/3\delta_{ij}\delta_{kl}$, is the volumetric part of the fourth-order unity tensor $\mathbb{I}$, and $\mathbb{K} = \mathbb{I} - \mathbb{J}$ is its deviatoric part. The Hill tensors are related to Eshelby tensors via Eq. (C.11). The Eshelby tensor $\mathbb{S}^{esh}_{sph}$ corresponding to spherical inclusions (pores in Fig. C.2) reads as

$$\mathbb{S}^{esh}_{sph} = \frac{3k_{poly}}{3k_{poly} + 4\mu_{poly}}\mathbb{J} + \frac{6(k_{poly} + 2\mu_{poly})}{5(3k_{poly} + 4\mu_{poly})}\mathbb{K} \quad \text{(C.16)}$$

In the base frame ($\boldsymbol{e}_\vartheta$, $\boldsymbol{e}_\varphi$, $\boldsymbol{e}_r$)(1=ϑ, 2=φ, 3=r, see Fig. C.3 for Euler angles φ and ϑ), attached to individual solid needles, the non-zero components of the Eshelby tensor $\mathbb{S}^{esh}_{cyl}$ corresponding to cylindrical inclusions read as

$$\begin{aligned} S^{esh}_{cyl,1111} = S^{esh}_{cyl,2222} &= \frac{5 - 4\nu_{poly}}{8(1-\nu_{poly})} \\ S^{esh}_{cyl,1122} = S^{esh}_{cyl,2211} &= \frac{-1 + 4\nu_{poly}}{8(1-\nu_{poly})} \\ S^{esh}_{cyl,1133} = S^{esh}_{cyl,2233} &= \frac{\nu_{poly}}{2(1-\nu_{poly})} \\ S^{esh}_{cyl,2323} = S^{esh}_{cyl,3232} = S^{esh}_{cyl,3223} = S^{esh}_{cyl,2332} &= \\ = S^{esh}_{cyl,3131} = S^{esh}_{cyl,1313} = S^{esh}_{cyl,1331} = S^{esh}_{cyl,3113} &= \frac{1}{4} \\ S^{esh}_{cyl,1212} = S^{esh}_{cyl,2121} = S^{esh}_{cyl,2112} = S^{esh}_{cyl,1221} &= \frac{3 - 4\nu_{poly}}{8(1-\nu_{poly})} \end{aligned} \quad \text{(C.17)}$$

with ν_{poly} as Poisson's ratio of the polycrystal,

$$\nu_{poly} = \frac{3k_{poly} - 2\mu_{poly}}{6k_{poly} + 2\mu_{poly}} \tag{C.18}$$

Following standard tensor calculus (Salencon 2001), the tensor components of $\mathbb{P}^{poly}_{cyl}(\vartheta, \varphi) = \mathbb{S}^{esh}_{cyl}(\vartheta, \varphi) : \mathbb{C}^{-1}_{poly}$, being related to differently oriented inclusions, are transformed into one, single base frame ($\boldsymbol{e}_1$, $\boldsymbol{e}_2$, $\boldsymbol{e}_3$), in order to evaluate the integrals in Eqs. (C.14) and (C.15).

C.3.2 Strength estimate

Strength of the porous polycrystal made up of hydroxyapatite needles (see Fig. C.2 for its RVE) is related to brittle failure of the most unfavorably stressed single needle. Therefore, the macroscopic stress (and strain) state needs to be related to corresponding stress and strain states in the individual needles. Accordingly, we specify the concentration relations (C.7) and (C.10) for the biomaterial defined through Eqs. (C.14)-(C.18), resulting in

$$\boldsymbol{\varepsilon}_{HA}(\varphi, \vartheta) = \left[\mathbb{I} + \mathbb{P}^{poly}_{cyl}(\varphi, \vartheta) : (\mathbb{c}_{HA} - \mathbb{C}_{poly})\right]^{-1} :$$

$$\left\{(1-\phi)\left\langle\left[\mathbb{I} + \mathbb{P}^{poly}_{cyl}(\varphi, \vartheta) : (\mathbb{c}_{HA} - \mathbb{C}_{poly})\right]^{-1}\right\rangle + \phi(\mathbb{I} - \mathbb{P}^{poly}_{sph} : \mathbb{C}_{poly})^{-1}\right\}^{-1} : \boldsymbol{E} \tag{C.19}$$

When employing phase elasticity (C.1) to hydroxyapatite, and overall elasticity (C.6) to the porous biomaterial according to Eq. (C.14), concentration relation (C.19) can be recast in terms of stresses

$$\boldsymbol{\sigma}_{HA}(\varphi, \vartheta) = \mathbb{c}_{HA} : \left\{\left[\mathbb{I} + \mathbb{P}^{poly}_{cyl}(\varphi, \vartheta) : (\mathbb{c}_{HA} - \mathbb{C}_{poly})\right]^{-1} :\right.$$

$$\left.\left\{(1-\phi)\left\langle\left[\mathbb{I} + \mathbb{P}^{poly}_{cyl}(\varphi, \vartheta) : (\mathbb{c}_{HA} - \mathbb{C}_{poly})\right]^{-1}\right\rangle + \phi(\mathbb{I} - \mathbb{P}^{poly}_{sph} : \mathbb{C}_{poly})^{-1}\right\}^{-1}\right\} :$$

$$\mathbb{C}^{-1}_{poly} : \boldsymbol{\Sigma} = \mathbb{B}_{HA}(\varphi, \vartheta) : \boldsymbol{\Sigma} \tag{C.20}$$

with $\mathbb{B}_{HA}(\varphi, \vartheta)$ as the so-called stress concentration factor of needle with orientation $\boldsymbol{N}(\varphi, \vartheta)$. We consider that needle failure is governed by the normal stress $\sigma_{HA,NN}(\varphi, \vartheta)$ in needle direction and by the shear stress in planes orthogonal to the needle direction, $\sigma_{HA,Nn}(\varphi, \vartheta; \psi)$ (see Fig. C.3),

$$\sigma_{HA,NN}(\varphi, \vartheta) = \boldsymbol{N} \cdot \boldsymbol{\sigma}_{HA}(\varphi, \vartheta) \cdot \boldsymbol{N} \tag{C.21}$$

$$\sigma_{HA,Nn}(\varphi, \vartheta; \psi) = \boldsymbol{N} \cdot \boldsymbol{\sigma}_{HA}(\varphi, \vartheta) \cdot \boldsymbol{n}(\psi) \tag{C.22}$$

depending on the direction $\boldsymbol{n}$ orthogonal to $\boldsymbol{N}$, specified through angle ψ (Fig. C.3),

$$\boldsymbol{n} = \cos\psi\, \boldsymbol{e}_\vartheta + \sin\psi\, \boldsymbol{e}_\varphi \tag{C.23}$$

More specifically, the failure criterion for the single needle considers interaction between tensile strength $\sigma_{HA}^{ult,t}$ and shear strength $\sigma_{HA}^{ult,s}$, and it reads as

$$\vartheta = 0, \ldots, \pi, \; \psi = 0, \ldots, 2\pi :$$

$$\mathfrak{f}_{HA}(\boldsymbol{\sigma}) = \max_{\vartheta} \left(\beta \max_{\psi} |\sigma_{HA,Nn}| + \sigma_{HA,NN} \right) - \sigma_{HA}^{ult,t} = 0 \tag{C.24}$$

with $\beta = \sigma_{HA}^{ult,t} / \sigma_{HA}^{ult,s}$ being the ratio between uniaxial tensile strength $\sigma_{HA}^{ult,t}$, and the shear strength $\sigma_{HA}^{ult,s}$ of pure hydroxyapatite. Use of Eqs. (C.20) to (C.23) in Eq. (C.24) yields a macroscopic failure criterion in the format of Eq. (C.13),

$$\mathfrak{F}(\boldsymbol{\Sigma}) = \max_{\vartheta} \left\{ \beta \max_{\psi} |\boldsymbol{N} \cdot \mathbb{B}_{HA}(\varphi, \vartheta) : \boldsymbol{\Sigma} \cdot \boldsymbol{n}(\psi)| + \boldsymbol{N} \cdot \mathbb{B}_{HA}(\varphi, \vartheta) : \boldsymbol{\Sigma} \cdot \boldsymbol{N} \right) - \sigma_{HA}^{ult,t} = 0 \tag{C.25}$$

and a corresponding elastic domain,

$$\mathfrak{F}(\boldsymbol{\Sigma}) < 0 \tag{C.26}$$

with $\mathbb{B}_{HA}$ according to Eq. (C.20). We also will evaluate the criterion (C.25) for uniaxial macroscopic stress states $\boldsymbol{\Sigma} = \pm \Sigma_{ref} \boldsymbol{e}_3 \otimes \boldsymbol{e}_3$: Insertion of these stress states into Eqs. (C.20)-(C.24) yields an equation for Σ_{ref}, the corresponding results $\Sigma_{poly}^{ult,t}$ and $\Sigma_{poly}^{ult,c}$ being model predictions of macroscopic uniaxial strengths as functions of (microscopic) needle strength and porosity (see Figs. C.6 and C.7, and Section C.4 for further details).

C.4 Model validation

C.4.1 Strategy for model validation through independent test data

In the line of Popper, who stated that a theory – as long as it has not been falsified – will be 'the more satisfactory the greater the severity of independent tests it survives' (cited from (Mayr 1997), p.49), the verification of the micromechanical representation of hydroxyapatite biomaterials [Eqs. (C.14)-(C.18) for elasticity, and Eqs. (C.19)-(C.26) for strength] will rest on two independent experimental sets, as it has been successfully done for other material classes such as bone (Hellmich and Ulm 2002b; Hellmich et al. 2004a; Fritsch and Hellmich 2007) or wood (Hofstetter et al. 2005, 2006). Biomaterial-specific macroscopic (homogenized) stiffnesses $\mathbb{C}_{poly}$ (Young's moduli E_{poly} and Poisson's ratios ν_{poly}), and uniaxial (tensile and compressive) strengths ($\Sigma_{poly}^{ult,t}$ and $\Sigma_{poly}^{ult,c}$), predicted by the micromechanics model (C.14)-(C.26) on the basis of biomaterial-independent (universal) elastic and strength properties of pure hydroxyapatite (experimental set I, Table C.2) for biomaterial-specific porosities ϕ (experimental set IIa, Tables C.3 and C.4), are compared to corresponding biomaterial-specific experimentally determined moduli E_{exp} and Poisson's ratios ν_{exp} (experimental set IIb-1, Table C.3) and uniaxial tensile/compressive strength values (experimental set IIb-2, Table C.4). Because we avoided introduction of micromorphological features that cannot be experimentally quantified (such as the precise crystal shape), all material parameters are directly related to well-defined experiments.

C.4.2 Universal mechanical properties of (biomaterial-independent) hydroxyapatite – Experimental set I

Experiments with an ultrasonic interferometer coupled with a solid media pressure apparatus (Katz and Ukraincik 1971; Gilmore and Katz 1982) reveal the isotropic elastic constants for dense hydroxyapatite powder ($\phi = 0$), the Young's modulus E_{HA}= 114 GPa, and the Poisson's ratio ν_{HA}= 0.27 (equivalent to bulk modulus $k_{HA} = E_{HA}/3/(1-2\nu_{HA})$= 82.6 GPa and shear modulus $\mu_{HA} = E_{HA}/2/(1+\nu_{HA})$= 44.9 GPa).

The authors are not aware of direct strength tests on pure hydroxyapatite (with $\phi = 0$). Therefore, we consider one uniaxial tensile test, $\Sigma_{exp}^{ult,t}$=37.1 MPa, and one uniaxial compressive test, $\Sigma_{exp}^{ult,c}$=509 MPa, on fairly dense samples (with ϕ=12.2% and ϕ=2.8%, respectively), conducted by Shareef et al. (1993) and Akao et al. (1981), respectively (see Table C.4). From these two tests, we back-calculate, via evaluation of Eqs. (C.20)-(C.25) for $\mathbf{\Sigma} = \Sigma_{exp}^{ult,t} \boldsymbol{e}_3 \otimes \boldsymbol{e}_3$ and $\mathbf{\Sigma} = -\Sigma_{exp}^{ult,c} \boldsymbol{e}_3 \otimes \boldsymbol{e}_3$, the universal tensile and shear strength of pure hydroxyapatite, $\sigma_{HA}^{ult,t}$ and $\sigma_{HA}^{ult,s}$ (Table C.2).

Young's modulus E_{HA}	114 GPa	from (Katz and Ukraincik 1971)
Poisson's ratio ν_{HA}	0.27	from (Katz and Ukraincik 1971)
Uniaxial tensile strength $\sigma_{HA}^{ult,t}$	52.2 MPa	from (Akao et al. 1981; Shareef et al. 1993);
Uniaxial shear strength $\sigma_{HA}^{ult,s}$	80.3 MPa	see Section C.4.2 for details

Table C.2: Universal (biomaterial-independent) isotropic phase properties of pure hydroxyapatite needles.

C.4.3 Biomaterial-specific porosities – Experimental set IIa

Porosity of hydroxyapatite biomaterials is standardly calculated from mass M and volume V of well-defined samples on the basis of the mass density of pure hydroxyapatite, ρ_{HA}=3.16 g/cm^3,

$$\phi = 1 - \frac{M}{V\,\rho_{HA}} \tag{C.27}$$

Corresponding porosity values have been reported by different investigators (Peelen et al. 1978; Akao et al. 1981; De With et al. 1981; Shareef et al. 1993; Arita et al. 1995; Martin and Brown 1995; Liu 1998; Charrière et al. 2001), see Tables C.3 and C.4.

Reference	ϕ (%)	E_{exp} (GPa)	ν_{exp} (1)
(Akao et al. 1981)	2.8	88	
	3.9	85	
	9.1	80	
	19.4	44	
(De With et al. 1981)	3	112	0.275
	6	103	0.272
	9	93	0.265
	17	78	0.253
	22	67	0.242
	27	54	0.238
(Arita et al. 1995)	6	88	
	28	41	
	33	32	
	35	29	
	50	14	
	52	10	
(Liu 1998)	8	93	
	17	78	
	21	66	
	32	44	
	44	22	
	54	18	
(Charrière et al. 2001)	44	13.5	

Table C.3: Experimental Young's modulus E_{exp} and Poisson's ratio ν_{exp} of hydroxyapatite biomaterials, as function of porosity ϕ.

C.4.4 Biomaterial-specific elasticity experiments on hydroxyapatite biomaterials – Experimental set IIb-1

Elastic properties of hydroxyapatite biomaterials were determined through uniaxial quasi-static mechanical tests (Akao et al. 1981; Charrière et al. 2001), but also through ultrasonic techniques (De With et al. 1981; Liu 1998), or resonance frequency tests (Arita et al. 1995).

In uniaxial quasi-static experiments, the gradient of the stress-strain curve gives access to Young's modulus. Respective experimental results are documented for cuboidal specimens (Akao et al. 1981) and hollow cylindrical specimens (Charrière et al. 2001), see Tables C.1 and C.3 as well as Fig. C.4.

In ultrasonic experiments (Ashman et al. 1984, 1987), the time of flight of an ultrasonic wave traveling through the specimen with a certain frequency f is measured. The calculated velocity of the wave, v, together with material mass density of the sample, gives access to the elastic constants (Carcione 2001; Kolsky 1953). Because the ultrasonic wavelength λ, $\lambda = v/f$, is a measure for the loading of the structure ($\lambda \approx L$ in Fig. C.1), the mechanical properties

Reference	ϕ (%)	$\Sigma_{exp}^{ult,c}$ (MPa)	$\Sigma_{exp}^{ult,t}$ (MPa)	$\Sigma_{exp}^{ult,b}$ (MPa)
(Peelen et al. 1978)	36	160		
	48	114		
	60	69		
	65	45		
	70	30		
(Akao et al. 1981)	2.8	509		
	3.9	465		
	9.1	415		
	19.4	308		
(Shareef et al. 1993)	12.2		37.1	
	16.1		32.8	
	20.6		31.8	
	24.8		24.2	
	27.3		23.6	
	29.2		20.0	
(Martin and Brown 1995)	27.0	172.5		
	39.0	119.0		
(Liu 1998)	20.2			25.5
	26.8			20.0
	29.0			16.8
	32.6			13.9
	39.6			14.4
	42.8			11.1
	50.9			7.2
	54.5			8.0

Table C.4: Experimental compressive strength $\Sigma_{exp}^{ult,c}$, bending strength $\Sigma_{exp}^{ult,b}$, and tensile strength $\Sigma_{exp}^{ult,t}$ of hydroxyapatite biomaterials, as functions of porosity ϕ.

are related to an RVE with characteristic length $l \ll \lambda$. Respective experimental results are documented for bar-shaped specimens (Liu 1998) and cylindrical samples (De With et al. 1981), see Tables C.1 and C.4 as well as Figs. C.4 and C.5.

In resonance frequency tests (Schreiber et al. 1973), beam type specimens are excited in the flexural vibration mode, and the corresponding free vibration gives access to the fundamental resonance frequency. The latter allows for determination, via the material mass density and the geometry of the sample, of the Young's modulus of the sample. Respective experimental results are documented for disc-shaped samples (Arita et al. 1995), see Tables C.1 and C.3 as well as Fig. C.4.

C.4.5 Comparison between biomaterial-specific stiffness predictions and corresponding experiments

The stiffness values predicted by the homogenization scheme (C.14)-(C.18) (see Section C.3 and Fig. C.2) for biomaterial-specific porosities (Section C.4.3, experimental set IIa) on the basis of biomaterial-independent (universal) stiffness of hydroxyapatite (Section C.4.2, experimental set I) are compared to corresponding experimentally determined biomaterial-specific stiffness values from experimental set IIb-1 (Section C.4.4). To quantify the model's predictive capabilities we consider the mean and the standard deviation of the relative error between stiffness predictions and experiments,

$$\bar{e} = \frac{1}{n}\sum e_i = \frac{1}{n}\sum \frac{q_{poly} - q_{exp}}{q_{exp}} \tag{C.28}$$

$$e_S = \left[\frac{1}{n-1}\sum (e_i - \bar{e})^2\right]^{\frac{1}{2}} \tag{C.29}$$

where q has to be replaced by the quantity in question, E or ν, and with summation over n stiffness values (see Tables C.3 and C.4).

Insertion of biomaterial-specific porosities (Table C.3) into Eq. (C.14) delivers, together with Eqs. (C.15) to (C.18), the biomaterial-specific stiffness estimates for the effective Young's modulus E_{poly} and the effective Poisson's ratio ν_{poly}. These stiffness predictions are compared to corresponding experimental stiffness values (Figs. C.4 and C.5). The satisfactory agreement between model predictions and experiments is quantified by prediction errors of 16±25% [mean value±standard deviation according to Eqs. (C.28) and (C.29)] for Young's modulus, and of -0.4±2.3% for Poisson's ratio.

C.4.6 Biomaterial-specific strength experiments on hydroxyapatite biomaterials – Experimental set IIb-2

In uniaxial compressive quasi-static tests, a sharp decrease of stress after a stress peak in the stress-strain diagram (Akao et al. 1981; Martin and Brown 1995) indicates brittle material failure, as observed for all biomaterials described herein, and the aforementioned stress peak is referred to as the ultimate stress or uniaxial strength $\Sigma_{exp}^{ult,c}$. Respective experimental results are documented for cylindrical samples (Peelen et al. 1978) and bars (Akao et al. 1981), see Tables C.1 and C.4 as well as Fig. C.7.

In three-point bending tests, a force F_s is applied to the centre of a beam specimen, and the maximum normal stress $\mathbf{\Sigma}^{ult} = \Sigma^{ult}\boldsymbol{e}_3 \otimes \boldsymbol{e}_3$ in the bar-type sample is calculated according to beam theory,

$$\Sigma_{exp}^{ult,t} = \frac{3F_s l_s}{2b_s h_s^2} \tag{C.30}$$

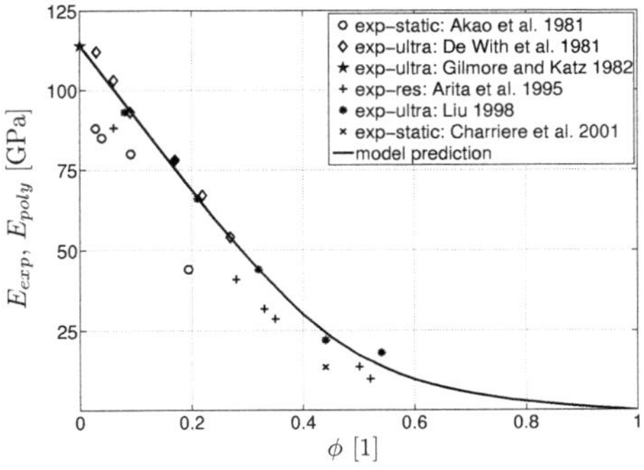

Figure C.4: Comparison between model predictions (E_{poly}) [Eqs. (C.14)-(C.18)] and experiments (E_{exp}) for Young's modulus of different porous biomaterials made of hydroxyapatite, as a function of porosity ϕ; ultra... ultrasonic tests, res ... resonance frequency tests, static... quasi-static uniaxial tests.

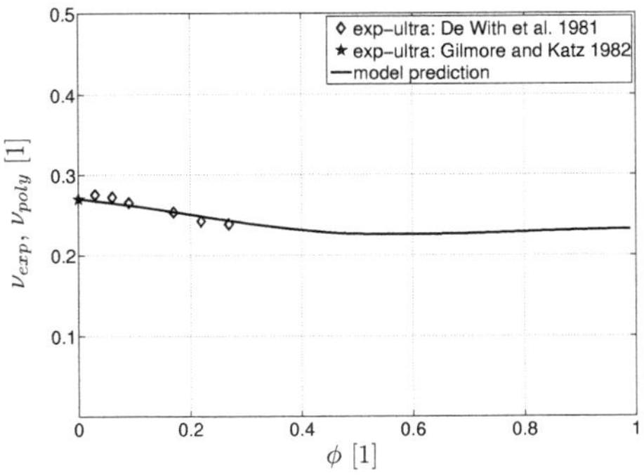

Figure C.5: Comparison between model predictions (ν_{poly}) [Eqs. (C.14)-(C.18)] and experiments (ν_{exp}) for Poisson's ratio of different porous biomaterials made of hydroxyapatite, as a function of porosity ϕ; ultra... ultrasonic tests.

with l_s, b_s, and h_s as the length, width and height of the specimen with rectangular cross-section, respectively. Respective experimental results (Liu 1998) are depicted in Tables C.1 and C.4 (there, bending strengths are denoted as $\Sigma_{exp}^{ult,b}$) as well as in Fig. C.6.

In the Stanford ring bursting test, ring-shaped specimens are pressurized internally, in order to generate a tensile hoop stress in the ring. The pressure is increased until the sample fails.

The tensile stress in the ring is calculated according to

$$\Sigma_{exp}^{ult,t} = \frac{r_s p_i}{d_s} \tag{C.31}$$

with r_s as the inner diameter of the ring, p_i as the internal pressure, and d_s as the wall thickness of the ring. Respective experimental results (Shareef et al. 1993) are depicted in Tables C.1 and C.4 as well as Fig. C.6.

C.4.7 Comparison between biomaterial-specific strength predictions and corresponding experiments

The strength values predicted by the homogenization scheme (C.19)-(C.26) (see Section C.3 and Fig. C.2) for biomaterial-specific porosities (Section C.4.3, experimental set IIa) on the basis of biomaterial-independent (universal) uniaxial tensile and compressive strengths of hydroxyapatite (Section C.4.2, experimental set I) are compared to corresponding experimentally determined biomaterial-specific uniaxial tensile and compressive strength values from experimental set IIb-2 (Section C.4.6).

Insertion of biomaterial-specific porosities (Table C.4) into Eqs. (C.14)-(C.25) delivers, together with E_{HA}, ν_{HA}, $\sigma_{HA}^{ult,t}$, and $\sigma_{HA}^{ult,s}$ (Table C.2), biomaterial-specific strength estimates for uniaxial tensile strength ($\Sigma_{poly}^{ult,t}$) and uniaxial compressive strength ($\Sigma_{poly}^{ult,c}$). These strength predictions are compared to corresponding experimental strength values (Figs. C.6 and C.7). The satisfactory agreement between model predictions and experiments is quantified by prediction errors of 14±15% for uniaxial tensile strength and -21±28% for uniaxial compressive strength; according to Eqs. (C.28) and (C.29) with $q_{poly} = \Sigma_{poly}^{ult,t}$ and $\Sigma_{poly}^{ult,c}$, respectively, and with $q_{exp} = \Sigma_{exp}^{ult,t}$ and $\Sigma_{exp}^{ult,c}$, respectively.

It is interesting to evaluate which crystal (located through the critical crystal angle ϑ_{cr} measured from the axis of macroscopic uniaxial loading) initiates the overall brittle material failure, and to find out at which crystal stresses this occurs (Figs. C.8 and C.9). Under tensile uniaxial macroscopic loading, failure occurs in crystals oriented closely to the loading direction (Fig. C.8), for the entire range of biomaterial porosities. In contrast, compressive uniaxial macroscopic loading induces failure in crystals which are oriented more or less perpendicularly to the loading direction, again for the entire range of biomaterial porosities. This is consistent with earlier findings that tensile loading leads to cracking perpendicular to the loading direction (mode I cracks) (Pichler et al. 2007b), and that compressive loading leads to cracks in the planes incorporating the load axis (axial splitting) (Pichler et al. 2007a). As regards the crystal stresses at failure, normal tensile stresses in needle direction prevail under tensile macroscopic loading, while tensile or compressive normal stresses combined with shear occur under compressive loading (Fig. C.9).

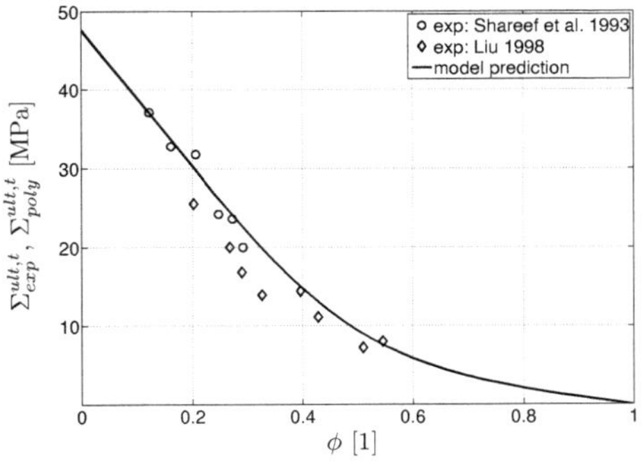

Figure C.6: Comparison between model predictions [Eqs. (C.14)-(C.25)] and experiments for tensile strength of different porous biomaterials made of hydroxyapatite, as a function of porosity ϕ.

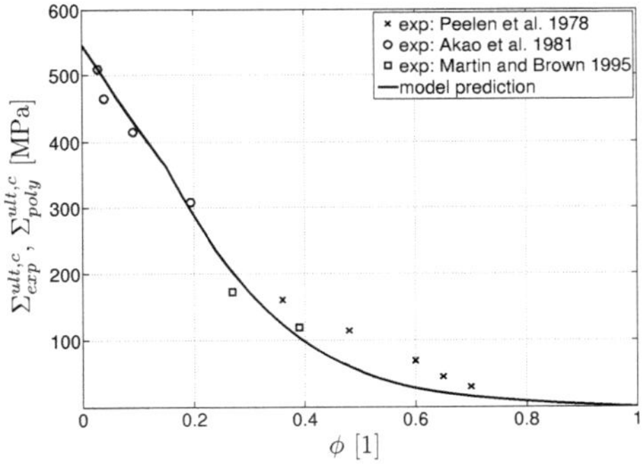

Figure C.7: Comparison between model predictions [Eqs. (C.14)-(C.25)] and experiments for compressive strength of different porous biomaterials made of hydroxyapatite, as a function of porosity ϕ.

C.5 Discussion

We have developed a continuum micromechanical concept for elasticity and strength of porous biomaterials made of hydroxyapatite, which was verified through independent experimental sets. We propose that such models have a considerable potential for improving biomaterial design. Nowadays the latter is largely done in a trial-and-error procedure. Based on a number of mechanical and/or acoustical tests, new material design parameters are guessed. On the other hand, with well validated micromechanics models, the mechanical implications of changes

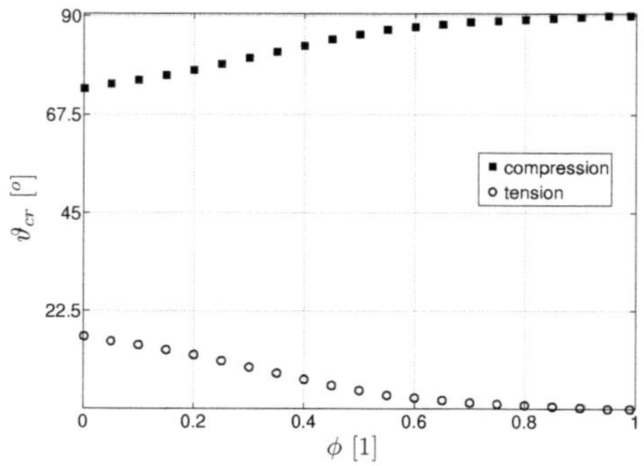

Figure C.8: Orientation of crystal needle initiating overall failure by fulfilling local failure criterion (C.24), measured through critical angle ϑ_{cr} from the loading direction, for tensile and compressive uniaxial macroscopic loading, as function of porosity ϕ.

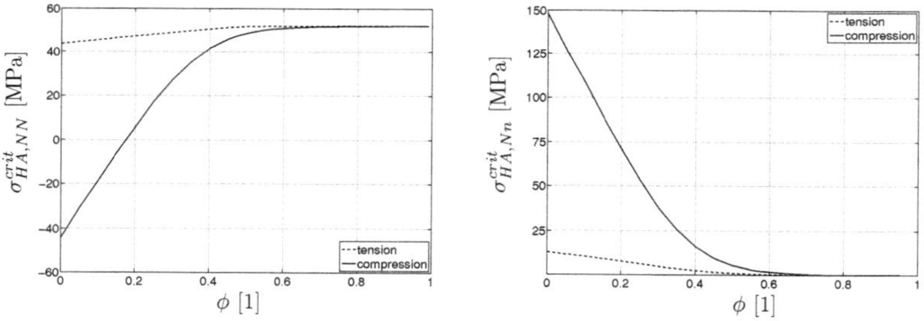

Figure C.9: Stress state in crystal needle fulfilling local failure criterion (C.24), in terms of (a) normal stresses and of (b) shear stresses in planes perpendicular to the needle direction, for tensile and compressive uniaxial macroscopic loading, as function of porosity ϕ.

in the microstructure can be predicted so that minimization of material failure risk allows for optimization of key design parameters, such as porosities or geometries of microstructures. Hence, we believe that micromechanical theories can considerably speed up the future improvement of tissue engineering scaffolds. In this context, extension of our modeling approach towards hydroxyapatite biomaterials with a hierarchical structure, i.e. with a double-porosity with different pore sizes (Woodard et al. 2007), and/or towards collagen/hydroxyapatite or chitosan/hydroxyapatite composite materials (Yunoki et al. 2006; Salgado et al. 2004) is currently under way.

C.6 Appendix: Nomenclature

$\mathbb{A}_r$	fourth-order strain concentration tensor of phase r
$\mathbb{A}_r^{est}$	estimate of fourth-order strain concentration tensor of phase r
$\mathbb{B}_{HA}$	fourth-order stress concentration tensor for single HA crystals
$\mathbb{C}$	fourth-order homogenized stiffness tensor
$\mathbb{C}^{est}$	estimate of fourth-order homogenized stiffness tensor
$\mathbb{C}^0$	fourth-order stiffness tensor of infinite matrix surrounding an ellipsoidal inclusion
$\mathbb{C}_{poly}$	fourth-order homogenized stiffness tensor of biomaterial made of HA
$\mathbb{c}_{HA}$	fourth-order stiffness tensor of single HA crystals within the RVE V_{poly}
$\mathbb{c}_r$	fourth-order stiffness tensor of phase r
d	characteristic length of inhomogeneity within an RVE
$\boldsymbol{E}$	second-order macroscopic strain tensor
E_{exp}	experimental Youngs modulus of biomaterial made of HA
E_{HA}	Youngs modulus of single HA crystals within the RVE V_{poly}
E_{poly}	homogenized Young's modulus of biomaterial made of HA
$\bar{e}$	mean of relative error between predictions and experiments
e_S	standard deviation of relative error between predictions and experiments
$\boldsymbol{e}_1$, $\boldsymbol{e}_2$, $\boldsymbol{e}_3$	unit base vectors of Cartesian reference base frame
$\boldsymbol{e}_\vartheta$, $\boldsymbol{e}_\varphi$, $\boldsymbol{e}_r$	unit base vectors of Cartesian local base frame of a single crystal
$\mathfrak{F}(\boldsymbol{\Sigma})$	boundary of elastic domain in space of macrostresses
f	ultrasonic excitation frequency
f_r	volume fraction of phase r
$\mathfrak{f}_r(\boldsymbol{\sigma})$	boundary of elastic domain of phase r in space of microstresses
HA	hydroxyapatite
$\mathbb{I}$	fourth-order identity tensor
$\mathbb{J}$	volumetric part of fourth-order identity tensor $\mathbb{I}$
$\mathbb{K}$	deviatoric part of fourth-order identity tensor $\mathbb{I}$
k_{HA}	bulk modulus of single HA crystals within the RVE V_{poly}
k_{poly}	homogenized bulk modulus of biomaterial made of HA
L	characteristic length of a structure built up by material RVEs
ℓ	characteristic length of RVEs
M	mass of a HA biomaterial sample
$\boldsymbol{N}$	orientation vector aligned with longitudinal axis of needle
N_r	number of phases within an RVE
$\boldsymbol{n}$	orientation vector perpendicular to $\boldsymbol{N}$
$\mathbb{P}_r^0$	fourth-order Hill tensor characterizing the interaction between the phase r and the matrix $\mathbb{C}^0$
$\mathbb{P}_{cyl}^{poly}$	fourth-order Hill tensor for cylindrical inclusion in matrix with stiffness $\mathbb{C}_{poly}$

$\mathbb{P}^{poly}_{sph}$	fourth-order Hill tensor for spherical inclusion in matrix with stiffness $\mathbb{C}_{poly}$
RVE	representative volume element
r, s	index for phases
$\mathbb{S}^{esh,0}_{r}$	fourth-order Eshelby tensor for phase r embedded in matrix $\mathbb{C}^0$
$\mathbb{S}^{esh}_{cyl}$	fourth-order Eshelby tensor for cylindrical inclusion embedded in isotropic matrix with stiffness $\mathbb{C}_{poly}$
$\mathbb{S}^{esh}_{sph}$	fourth-order Eshelby tensor for spherical inclusion embedded in isotropic matrix with stiffness $\mathbb{C}_{poly}$
tr	trace of a second-order tensor
V	volume of a HA biomaterial sample
V_{RVE}	volume of an RVE
v	ultrasonic wave propagation velocity within a HA biomaterial sample
w	index denoting weakest phase
$\boldsymbol{x}$	position vector within an RVE
β	ratio between uniaxial tensile strength and shear strength of pure HA
δ_{ij}	Kronecker delta (components of second-order identity tensor $\mathbf{1}$)
ε_{HA}	second-order strain tensor field within single HA crystals
ε_r	second-order strain tensor field of phase r
ϑ	latitudinal coordinate of spherical coordinate system
λ	ultrasonic wave length
μ_{HA}	shear modulus of single HA crystals within the RVE V_{poly}
μ_{poly}	homogenized shear modulus of biomaterial made of HA
ν_{exp}	experimental Poisson's ratio of biomaterial made of HA
ν_{HA}	Poisson's ratio of single HA crystals within the RVE V_{poly}
ν_{poly}	homogenized Poisson's ratio of biomaterial made of HA
$\boldsymbol{\xi}$	displacements within an RVE and at its boundary
ρ	material mass density
ρ_{HA}	mass density of pure HA
ρ_s	mass density of a HA biomaterial sample
$\boldsymbol{\Sigma}$	second-order macroscopic stress tensor
$\Sigma^{ult,t}_{poly}$	model-predicted uniaxial tensile strength of biomaterial made of HA
$\Sigma^{ult,c}_{poly}$	model-predicted uniaxial compressive strength of biomaterial made of HA
$\Sigma^{ult,t}_{exp}$	experimental uniaxial tensile strength of biomaterial made of HA
$\Sigma^{ult,c}_{exp}$	experimental uniaxial compressive strength of biomaterial made of HA
Σ_{ref}	component of uniaxial stress tensor $\boldsymbol{\Sigma}$ imposed on boundary of biomaterial made of HA
$\boldsymbol{\sigma}_{HA}(\varphi,\vartheta)$	second-order stress tensor field within single HA crystals
$\sigma_{HA,NN}(\varphi,\vartheta)$	normal component of stress tensor $\boldsymbol{\sigma}_{HA}(\varphi,\vartheta)$ in needle direction
$\sigma_{HA,Nn}(\varphi,\vartheta)$	shear component of stress tensor $\boldsymbol{\sigma}_{HA}(\varphi,\vartheta)$ in planes orthogonal to the needle direction

$\sigma_{HA}^{ult,t}$	uniaxial tensile strength of pure HA
$\sigma_{HA}^{ult,s}$	shear strength of pure HA
$\boldsymbol{\sigma}_r$	second-order stress tensor field of phase r
φ	longitudinal coordinate of spherical coordinate system
ϕ	volume fraction of micropores within RVE of porous HA
ψ	longitudinal coordinate of vector $\boldsymbol{n}$
∂V	boundary of an RVE
$\mathbf{1}$	second-order identity tensor
$\langle(.)\rangle_V = 1/V \int_V (.)\mathrm{d}V$	average of quantity (.) over volume V
$\cdot$	first-order tensor contraction
$:$	second-order tensor contraction
$\otimes$	dyadic product of tensors

Publication D

Ductile sliding between mineral crystals followed by rupture of collagen crosslinks: experimentally supported micromechanical explanation of bone strength (Fritsch et al. 2009b)

Authored by Andreas Fritsch, Christian Hellmich, and Luc Dormieux
Submitted for publication to *Journal of Theoretical Biology*

There is an ongoing discussion on how bone strength could be explained from its internal structure and composition. Reviewing recent experimental and molecular dynamics studies, we here propose a new vision on bone material failure: mutual ductile sliding of hydroxyapatite mineral crystals along layered water films is followed by rupture of collagen crosslinks. In order to cast this vision into a mathematical form, a multiscale continuum micromechanics theory for upscaling of elastoplastic properties is developed, based on the concept of concentration and influence tensors for eigenstressed microheterogeneous materials. The model reflects bone's hierarchical organization, in terms of representative volume elements for cortical bone, for extravascular and extracellular bone material, for mineralized fibrils and the extrafibrillar space, and for wet collagen. In order to get access to the stress states at the interfaces between crystals, the extrafibrillar mineral is resolved into an infinite amount of cylindrical material phases oriented in all directions in space. The multiscale micromechanics model is shown to be able to satisfactorily predict the strength characteristics of different bones from different species, on the basis of their mineral/collagen content, their intercrystalline, intermolecular, lacunar, and vascular porosities, and the elastic and strength properties of hydroxyapatite and (molecular) collagen.

D.1 Introduction

Explanation of the highly diverse mechanical properties of the material bone from its internal structure and composition has been a biomechanician's wish (Fung 2002; Martin et al. 1998), ever since the establishment of this scientific field. This wish has motivated (i) comprehensive mechanical testing series over all types of tissues and vertebrates (led by Currey and colleagues (Currey 1959; Reilly and Burstein 1974b; Keaveny et al. 1993)), (ii) the incorporation of the theory of anisotropic elasticity in the framework of ultrasonic testing (driven forward by Katz and colleagues (Katz 1980; Ashman et al. 1984)), and (iii) the complementation of the aforementioned two activities with chemical and physical measurements revealing micro and nanostructural features of mineralized collagenous tissues (pioneered in an unparalleled experimental campaign by Lees and colleagues (Lees et al. 1979b,a, 1983; Lees 1987a)). The huge experimental legacy following from the aforementioned activities was theoretically integrated in the context of validating micromechanical models holding for bone materials across different species, ages and anatomical locations (Hellmich and Ulm 2002a; Hellmich et al. 2004a; Hellmich and Ulm 2005a; Fritsch and Hellmich 2007). Such micromechanical models predict, on the basis of mechanical properties of bone elementary constituents (hydroxyapatite, collagen, water), the (poro-) elasticity tensors at the different hierarchical levels of the material, from tissue-specific composition data, such as porosities and mineral/collagen content. Therefore, morphological features such as Haversian and lacunar, intercrystalline, and intermolecular porosities, mineralized fibrils and collagen-free extrafibrillar space, plate or needle-type hydroxyapatite crystals and long crosslinked collagen molecules were represented in the framework of continuum micromechanics, also referred to as random homogenization theory (Hill 1963; Suquet 1997b; Zaoui 2002). A key feature of these micromechanical models is the explicit consideration of the extrafibrillar mineral crystals whose existence was evidenced earlier (Lees et al. 1984a, 1994; Prostak and Lees 1996; Pidaparti et al. 1996; Benezra Rosen et al. 2002). In this sense, the challenge of micromechanics-supported, consistently upscaled microstructure-property relationships for poroelasticity in bone has been met quite reasonably.

However, the case of explaining bone strength from its internal structure and composition seems to be fairly unsettled: while scaling relations for the strength of trabecular bone as function of porosity have become classical (Gibson 1985; Gibson and Ashby 1997), the micro and nanostructural origin of bone strength remains an open question: While several researchers favor the idea of brittle mineral crystals embedded in a compliant ductile organic (collagenous) matrix (Currey 1969; Katz 1980, 1981; Sasaki 1991; Mammone and Hudson 1993; Jäger and Fratzl 2000; Kotha and Guzelsu 2003) (still, explanation of a large number of experimental data through only one model and realistic prediction of measured stress-strain curves are somewhat out of sight), experiments show that collagen may actually fail in a quasi-brittle fashion (Christiansen et al. 2000; Gentleman et al. 2003), and this observation is confirmed by latest molecular dynamics simulations (Buehler 2006; Bhowmik et al. 2007). Such computations are essential tools for understanding the interaction of huge numbers of molecules, but, due to computational constraints, the largest models which can be realized nowadays are of the order of some hundreds of nanometers (Buehler 2006), far away from the larger length scales spanned by the material bone up to its macroscopic appearance at the millimeter to centimeter scale. What further complicates the matter is that once the elementary constituents mineral and collagen have failed, a complex series of crack propagation events starts, spanning length scales between tens of nanometers and ultimately several millimeters. Related toughening strategies in bone have been intensively studied (Burr et al. 1998; Reilly and Currey 2000; Akkus and Rimnac 2001; Okumura and Gennes 2001; Taylor et al. 2003; Ballarini et al. 2005; O'Brien et al. 2007; Koester et al. 2008), but a consistent mathematical theory for relating them to the overall, tissue-specific bone strength seems to be an enormously difficult task. Given this highly challenging situation, we ask: Can continuum micromechanics help to explain not only bone elasticity, but also bone strength from the material's internal structure and composition?

It is often felt that, in contrast to the elastic case, homogenization techniques which often refer to strains or stresses averaged over the material's constituents, might not help for the explanation of bone strength, where stress peaks are likely to govern material failure. Fortunately, this is not necessarily true: one remedy lies in the resolution of one material constituent into an infinite amount of sub-phases – e.g. the mineral phase may be split into an infinite amount of differently oriented needles, giving access to information on local stress peaks in these needles. It was recently shown (Fritsch et al. 2009a) that based on such a concept, the brittle failure of various hydroxyapatite biomaterials characterized by different porosities could be explained from the failure characteristics of individual crystals (quantified in terms of two strength values only) and from the microstructure these crystals build up.

This recent micromechanics model can deliver important input, in terms of the strength properties of single hydroxyapatite crystals, for a micromechanics model explaining bone strength – the latter is the focus of the present paper. It is organized as follows: Reviewing recent experimental and molecular dynamics studies, we first propose a new vision on bone material failure: mutual ductile sliding of mineral crystals along layered water films is followed by rupture of collagen crosslinks. In order to cast this vision into a mathematical form, we then present a continuum micromechanics theory for upscaling of elastoplastic properties. Thereafter, this theory is applied to a multiscale representation of bone materials. Conclusively, it is shown that the corresponding multiscale model can satisfactorily predict the stress-strain curves and the strength values of different bones from different species, on the basis of their mineral/collagen content, their intercrystalline, intermolecular, lacunar, and vascular porosities, and the elastic and strength properties of hydroxyapatite and collagen.

D.2 A new proposition for bone failure: layered water-induced ductile sliding of minerals, followed by rupture of collagen crosslinks

Classically, the strength of bone materials is thought to be related to the strength properties of collagen, to the strength properties of hydroxyapatite, and/or the interfaces between these constituents. However, more recent works extend and modify this traditional picture, by indicating the great role of water for the failure properties of bone. In this context, molecular dynamics studies on collagen molecules being detached from hydroxyapatite in solvated conditions, revealed that the interaction energies between hydroxyapatite and water, and between collagen and water, are by orders of magnitude larger than that between hydroxyapatite and collagen (Bhowmik et al. 2007). This implies that water probably plays a central role in 'glueing' together the material's elementary constituents, mineral with mineral, collagen with collagen, and also mineral with collagen. The latter interaction was confirmed by solid state Nuclear Magnetic Resonance (^{1}H NMR) studies (Wilson et al. 2006). As concerns the water-hydroxyapatite interactions, molecular dynamics simulations of crystal systems surrounded by water molecules revealed two to three well-organized water layers on the crystal surfaces, these structured water layers having ice-like features (Pan et al. 2007). These features were shown to chemically stabilize the crystals. In the present contribution, we will discuss the possibility that they also mechanically stabilize the interaction between mineral crystals: More specifically, we consider the case when the mineral crystals will not break or detach one from another once a critical stress threshold is reached (as in dry conditions), but when the intra- and intercrystalline loads accumulated up to the elastic limit, will be maintained through the (hydrated) crystals starting to glide upon each other, along the ice-like features, which prevent the sliding hydroxyapatite surfaces from disintegration. The latter is also prevented by the collagen fibrils interweaving the extracellular bone matrix. This vision is consistent with an elastoplastic interface behavior between hydrated hydroxyapatite. However, from a mathematical viewpoint, modeling interfaces between non-spherical objects is extremely expending (or extremely complex), so that we will benefit from the recent finding (Fritsch et al. 2009a) that the effect of 'micro'-interface behavior of elongated 1D particles, on the overall 'macroscopic' material can be mimicked by equivalent 'bulk' failure properties of the elongated phases. In case of hydroxyapatite polycrystals, we even know the (brittle) failure properties of the single hydroxyapatite crystals, and we will use them as elastic limits in the framework of full elastoplastic analysis of the hierarchical mineral-collagen-water composites called 'bone'. Therefore, it is appropriate to present a continuum micromechanics theory for elastoplasticity next. Thereby, our focus is on the plastic gliding mechanisms between mineral crystals, and we only proceed our computations until a critical stress in the collagen is reached. Potentially plastic behavior or microcracking events/crack bridging occuring thereafter (Nalla et al. 2004) are beyond our present scope. The critical stress of collagen is derived from direct mechanical

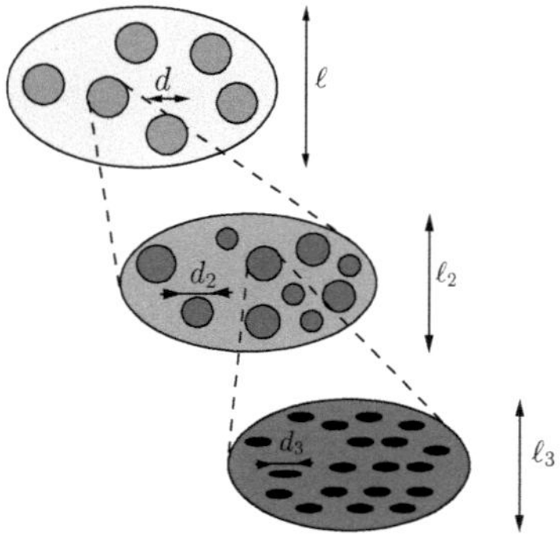

Figure D.1: Multistep homogenization: Properties of phases (with characteristic lengths of d and d_2, respectively) inside RVEs with characteristic lengths of ℓ or ℓ_2, respectively, are determined from homogenization over smaller RVEs with characteristic lengths of $\ell_2 \leq d$ and $\ell_3 \leq d_2$, respectively.

experiments on collagen, showing a brittle behavior of this constituent (Catanese et al. 1999; Christiansen et al. 2000; Gentleman et al. 2003), which is in agreement with some molecular dynamics studies (Buehler 2006, 2008; Vesentini et al. 2005). In particular, the latter work shows that collagen rupture is likely to be related to failure of crosslinks, such as the decorin molecule.

D.3 Fundamentals of continuum micromechanics – random homogenization of elastoplastic properties

D.3.1 Representative volume element

In continuum micromechanics (Hill 1963; Suquet 1997b; Zaoui 1997b, 2002), a material is understood as a macro-homogeneous, but micro-heterogeneous body filling a representative volume element (RVE) with characteristic length ℓ, $\ell \gg d$, d standing for the characteristic length of inhomogeneities within the RVE (see Fig. D.1), and $\ell \ll \mathcal{L}$, $\mathcal{L}$ standing for the characteristic lengths of geometry or loading of a structure built up by the material defined on the RVE. In general, the microstructure within one RVE is so complicated that it cannot be described in complete detail. Therefore, quasi-homogeneous subdomains with known physical quantities (such as volume fractions or elastoplastic properties) are reasonably chosen. They are called material phases. The 'homogenized' mechanical behavior of the overall material, i.e. the relation between homogeneous deformations acting on the boundary of the RVE and resulting (average) stresses, including the ultimate stresses sustainable by the RVE, can then be estimated from the mechanical behavior of the aforementioned homogeneous phases (representing the inhomogeneities within the RVE), their dosages within the RVE, their characteristic shapes, and their interactions. If a single phase exhibits a heterogeneous microstructure itself, its mechanical behavior can be estimated by introduction of an RVE within this phase, with dimensions $\ell_2 \leq d$, comprising again smaller phases with characteristic length $d_2 \ll \ell_2$, and so on, leading to a multistep homogenization scheme (see Fig. D.1).

D.3.2 Upscaling of elastoplastic properties

We consider an RVE consisting of n_r material phases, $r = 1, \ldots, n_r$, exhibiting elastoplastic material behavior, i.e. following the constitutive laws of ideal associated elastoplasticity,

$$\boldsymbol{\sigma}_r = \mathbb{c}_r : (\boldsymbol{\varepsilon}_r - \boldsymbol{\varepsilon}_r^p) \tag{D.1}$$

$$\dot{\boldsymbol{\varepsilon}}_r^p = \dot{\lambda}_r \frac{\partial \mathfrak{f}_r}{\partial \boldsymbol{\sigma}_r}, \quad \dot{\lambda}_r \mathfrak{f}_r(\boldsymbol{\sigma}_r) = 0, \quad \dot{\lambda}_r \geq 0, \quad \mathfrak{f}_r(\boldsymbol{\sigma}_r) \leq 0 \tag{D.2}$$

In Eq. (D.2), $\boldsymbol{\sigma}_r$ and $\boldsymbol{\varepsilon}_r$ are the stress and (linearized) strain tensors averaged over phase r with elasticity tensor $\mathbb{c}_r$; $\boldsymbol{\varepsilon}_r^p$ are the average plastic strains in phase r, λ_r is the plastic multiplier of phase r, and $\mathfrak{f}_r(\boldsymbol{\sigma}_r)$ is the yield function describing the (ideally) plastic characteristics of phase r. The RVE is subjected to Hashin boundary conditions, i.e. to 'homogeneous' ('macroscopic') strains $\boldsymbol{E}$ at its boundary, so that the kinematically compatible phase strains $\boldsymbol{\varepsilon}_r$ inside the RVE fulfill the average condition

$$\boldsymbol{E} = \sum_r f_r \boldsymbol{\varepsilon}_r \tag{D.3}$$

with f_r as the volume fraction of phase r. In a similar way, the equilibrated phase stresses $\boldsymbol{\sigma}_r$ fulfill the stress average condition

$$\boldsymbol{\Sigma} = \sum_r f_r \boldsymbol{\sigma}_r \tag{D.4}$$

with $\boldsymbol{\Sigma}$ as the 'macroscopic' stresses.

The superposition principle (following from linear elasticity and linearized strain) implies that the phase strains $\boldsymbol{\varepsilon}_r$ are linearly related to both the macroscopic strains $\boldsymbol{E}$, and to the free strains $\boldsymbol{\varepsilon}_r^p$ (which can be considered as independent loading parameters),

$$\boldsymbol{\varepsilon}_r = \mathbb{A}_r : \boldsymbol{E} + \sum_s \mathbb{d}_{rs} : \boldsymbol{\varepsilon}_s^p \tag{D.5}$$

with $\mathbb{A}_r$ as the fourth-order concentration tensor (Hill 1965), and $\mathbb{d}_{rs}$ as the fourth-order influence tensors (Dvorak 1992). The latter quantify the phase strains $\boldsymbol{\varepsilon}_r$ resulting from plastic strains $\boldsymbol{\varepsilon}_s^p$, while the overall RVE is free from deformation, $\boldsymbol{E} = \mathbf{0}$.

In absence of plastic strains [$\mathfrak{f}_r < 0$, $\boldsymbol{\varepsilon}_r^p = \mathbf{0}$ in Eqs. (D.1)-(D.2)], the RVE behaves fully elastically, so that (D.5), (D.4), (D.3), and (D.1) yield a macroscopic elastic law of the form

$$\boldsymbol{\Sigma} = \mathbb{C}^{hom} : \boldsymbol{E} \quad \text{with} \quad \mathbb{C}^{hom} = \sum_r f_r \mathbb{c}_r : \mathbb{A}_r \tag{D.6}$$

as the homogenized elastic stiffness tensor characterizing the material within the RVE. In case of non-zero 'free' plastic strains $\boldsymbol{\varepsilon}_r^p$, (D.6) can be extended to the form

$$\boldsymbol{\Sigma} = \mathbb{C}^{hom} : (\boldsymbol{E} - \boldsymbol{E}^p) \tag{D.7}$$

(D.7), together with (D.1), (D.4), (D.5), and (D.6) gives access to the macroscopic plastic strains $\boldsymbol{E}^p$, reading as

$$\boldsymbol{E}^p = -\left[\sum_r f_r \mathbb{c}_r : \mathbb{A}_r\right]^{-1} : \left\{\sum_r f_r \mathbb{c}_r : \left[(\mathbb{A}_r : \boldsymbol{E} + \sum_s \mathbb{a}_{rs} : \boldsymbol{\varepsilon}_s^p) - \boldsymbol{\varepsilon}_r^p\right]\right\} + \boldsymbol{E} \tag{D.8}$$

D.3.3 Matrix-inclusion based estimation of concentration and influence tensors

We estimate the concentration and influence tensors from matrix-inclusion problems, as it is standardly done in the field of elasticity homogenization. However, we consider not only elastic, but also free (plastic) strains in both the inclusion (with stiffness $\mathbb{c}_{inc}$) and surrounding infinite matrix (with stiffness $\mathbb{C}^0$); these plastic strains are denoted by $\boldsymbol{\varepsilon}_{inc}^p$ and $\boldsymbol{E}^{0,p}$. At its infinite boundary, the infinite matrix is subjected to homogeneous strains $\boldsymbol{E}^\infty$. Then, the strains in the inhomogeneity can be given in the form (Zaoui 2002)

$$\boldsymbol{\varepsilon}_{inc} = [\mathbb{I} + \mathbb{P}_{inc}^0 : (\mathbb{c}_{inc} - \mathbb{C}^0)]^{-1} : [\boldsymbol{E}^\infty + \mathbb{P}_{inc}^0 : (\mathbb{c}_{inc} : \boldsymbol{\varepsilon}_{inc}^p - \mathbb{C}^0 : \boldsymbol{E}^{0,p})] \tag{D.9}$$

We estimate the strains in phase r, $\boldsymbol{\varepsilon}_r$, as those of an inclusion of the same shape as the phase, i.e. we identify $\boldsymbol{\varepsilon}_{inc} = \boldsymbol{\varepsilon}_r$ in (D.9), and insert this result into the strain average rule (D.3), which yields a relation between $\boldsymbol{E}^\infty$ and $\boldsymbol{E}$,

$$\boldsymbol{E}^\infty = \left\{\sum_r f_r [\mathbb{I} + \mathbb{P}_r^0 : (\mathbb{c}_r - \mathbb{C}^0)]^{-1}\right\}^{-1} : \left\{\boldsymbol{E} - \sum_s f_s [\mathbb{I} + \mathbb{P}_s^0 : (\mathbb{c}_s - \mathbb{C}^0)]^{-1} : \mathbb{P}_s^0 : (\mathbb{c}_s : \boldsymbol{\varepsilon}_s^p - \mathbb{C}^0 : \boldsymbol{E}^{0,p})\right\} \tag{D.10}$$

Use of Eq. (D.10) in (D.9) specified for $\boldsymbol{\varepsilon} = \boldsymbol{\varepsilon}_r$ yields

$$\boldsymbol{\varepsilon}_r = [\mathbb{I} + \mathbb{P}_r^0 : (\mathbb{c}_r - \mathbb{C}^0)]^{-1} : \left\{\left\{\sum_i f_i [\mathbb{I} + \mathbb{P}_i^0 : (\mathbb{c}_i - \mathbb{C}^0)]^{-1}\right\}^{-1} : \left\{\boldsymbol{E} - \sum_s f_s [\mathbb{I} + \mathbb{P}_s^0 : (\mathbb{c}_s - \mathbb{C}^0)]^{-1} : \mathbb{P}_s^0 : (\mathbb{c}_s : \boldsymbol{\varepsilon}_s^p - \mathbb{C}^0 : \boldsymbol{E}^{0,p})\right\} + \mathbb{P}_r^0 : (\mathbb{c}_r : \boldsymbol{\varepsilon}_r^p - \mathbb{C}^0 : \boldsymbol{E}^{0,p})\right\} \tag{D.11}$$

In (D.11), the properties of the fictitious matrix, $\mathbb{C}^0$ and $\boldsymbol{E}^{0,p}$, still need to be chosen. As regards $\mathbb{C}^0$, its choice governs the interactions between the phases inside the RVE: $\mathbb{C}^0 = \mathbb{C}^{hom}$ relates to a dispersed arrangement of phases where all phases 'feel' the overall homogenized material, and the corresponding homogenization scheme is standardly called self-consistent (Hershey 1954; Kröner 1958), well-suited for polycrystalline materials. On the other hand, the matrix may be identified as a phase M itself, $\mathbb{C}^0 = \mathbb{c}_M$, which relates to a matrix-inclusion-type composite, and the corresponding homogenization scheme is standardly referred to as Mori-Tanaka scheme (Mori and Tanaka 1973; Benveniste 1987). Herein, we have to make an additional choice, relating to the plastic (free) strains in the fictitious matrix, $\boldsymbol{E}^{0,p}$. For a matrix-inclusion composite (Mori-Tanaka scheme), it seems natural to identify $\boldsymbol{E}^{0,p}$ with the free strain in the matrix phase, $\boldsymbol{\varepsilon}_M^p$. In case of the self-consistent scheme, however, we have to remember that the fictitious matrix does not exhibit any volume fractions – therefore, it cannot host any free strains, and $\boldsymbol{E}^{0,p}$ is set zero in that case. In particular, one is not allowed to set $\boldsymbol{E}^{0,p}$ equal to the macroscopic plastic strains prevailing at the RVE level, since this would be in conflict with the concentration relation (D.5).

Concentration relation (D.5) remains to be specified for the polycrystals and matrix-inclusion composites: For the former (self-consistent scheme, $\mathbb{C}^0 = \mathbb{C}^{hom}$, $\boldsymbol{E}^{0,p} = \mathbf{0}$), (D.11) reads as

$$\boldsymbol{\varepsilon}_r = [\mathbb{I} + \mathbb{P}_r^0 : (\mathbb{c}_r - \mathbb{C}^{hom})]^{-1} : \left\{ \left\{ \sum_i f_i [\mathbb{I} + \mathbb{P}_i^0 : (\mathbb{c}_i - \mathbb{C}^{hom})]^{-1} \right\}^{-1} : \right.$$

$$\left. \left\{ \boldsymbol{E} - \sum_s f_s [\mathbb{I} + \mathbb{P}_s^0 : (\mathbb{c}_s - \mathbb{C}^{hom})]^{-1} : \mathbb{P}_s^0 : \mathbb{c}_s : \boldsymbol{\varepsilon}_s^p \right\} + \mathbb{P}_r^0 : \mathbb{c}_r : \boldsymbol{\varepsilon}_r^p \right\} \tag{D.12}$$

Comparing (D.12) with (D.5), we can identify the concentration and influence tensors as

$$\mathbb{A}_r = \left[\mathbb{I} + \mathbb{P}_r^0 : (\mathbb{c}_r - \mathbb{C}^{hom})\right]^{-1} : \left\{ \sum_s f_s \left[\mathbb{I} + \mathbb{P}_s^0 : (\mathbb{c}_s - \mathbb{C}^{hom})\right]^{-1} \right\}^{-1} \tag{D.13}$$

and

$$\mathbb{D}_{rs} = \mathbb{D}_{rr} = (-f_r \mathbb{A}_r + \mathbb{I}) : (\mathbb{A}_r^\infty : \mathbb{P}_r^0 : \mathbb{c}_r) \quad \text{if} \quad r = s \tag{D.14}$$

otherwise

$$\mathbb{D}_{rs} = -f_s \mathbb{A}_r : \mathbb{A}_s^\infty : \mathbb{P}_s^0 : \mathbb{c}_s \tag{D.15}$$

whereby

$$\mathbb{A}_r^\infty = [\mathbb{I} + \mathbb{P}_r^0 : (\mathbb{c}_r - \mathbb{C}^{hom})]^{-1} \tag{D.16}$$

For the Mori-Tanaka case ($\mathbb{C}^0 = \mathbb{c}_M$, $\boldsymbol{E}^{0,p} = \boldsymbol{\varepsilon}^p_M$), (D.11) reads as

$$\boldsymbol{\varepsilon}_r = [\mathbb{I} + \mathbb{P}^0_r : (\mathbb{c}_r - \mathbb{c}_M)]^{-1} : \left\{ \left\{ \sum_i f_i [\mathbb{I} + \mathbb{P}^0_i : (\mathbb{c}_i - \mathbb{c}_M)]^{-1} \right\}^{-1} : \right.$$
$$\left\{ \boldsymbol{E} - \sum_s f_s [\mathbb{I} + \mathbb{P}^0_s : (\mathbb{c}_s - \mathbb{c}_M)]^{-1} : [\mathbb{P}^0_s : (\mathbb{c}_s : \boldsymbol{\varepsilon}^p_s - \mathbb{c}_M : \boldsymbol{\varepsilon}^p_M)] \right\}$$
$$\left. + \mathbb{P}^0_r : (\mathbb{c}_r : \boldsymbol{\varepsilon}^p_r - \mathbb{c}_M : \boldsymbol{\varepsilon}^p_M) \right\} \tag{D.17}$$

Comparing (D.17) with (D.5), we can identify the concentration and influence tensors as

$$\mathbb{A}_r = \left[\mathbb{I} + \mathbb{P}^0_r : (\mathbb{c}_r - \mathbb{c}_M)\right]^{-1} : \left\{ \sum_s f_s \left[\mathbb{I} + \mathbb{P}^0_s : (\mathbb{c}_s - \mathbb{c}_M)\right]^{-1} \right\}^{-1} \tag{D.18}$$

and

$$\mathbb{a}_{rs} = \mathbb{a}_{rr} = (-f_r \mathbb{A}_r + \mathbb{I}) : (\mathbb{A}^\infty_r : \mathbb{P}^0_r : \mathbb{c}_r) \qquad \text{if} \quad r = s \tag{D.19}$$

$$\mathbb{a}_{rs} = \mathbb{a}_{rM} = \mathbb{A}_r : (-f_M \mathbb{A}^\infty_M : \mathbb{P}^0_M : \mathbb{c}_M +$$
$$\sum_i f_i \mathbb{A}^\infty_i : \mathbb{P}^0_i : \mathbb{c}_M) - -\mathbb{A}^\infty_r : \mathbb{P}^0_r : \mathbb{c}_M \qquad \text{if} \quad s = M \tag{D.20}$$

otherwise

$$\mathbb{a}_{rs} = -f_s \mathbb{A}_r : \mathbb{A}^\infty_s : \mathbb{P}^0_s : \mathbb{c}_s \tag{D.21}$$

D.4 Application of microelastoplastic theory to bone

In the following, we will apply the above developed microelastoplastic theory to the material 'cortical bone'. Therefore, we will employ a slight adaptation of a recently proposed and validated multiscale material model for bone elasticity (Fritsch and Hellmich 2007), see Figure D.2. The adaptation lies in considering different orientations of non-spherical mineral crystals, as this precision of morphological resolution is mandatory for the appropriate prediction of the material's strength properties, as has been shown for other materials such as hydroxyapatite biomaterials (Fritsch et al. 2009a), concrete (Pichler et al. 2008a,b), or gypsum (Sanahuja et al. 2008). As the basis for such a multiscale micromechanics model, the mechanical properties of the elementary components, of hydroxyapatite, of collagen, and of water, are required. They will be discussed first.

D.4.1 Elastic properties of hydroxyapatite, collagen, and water

Concerning the tissue-independent ('universal') phase properties of the elementary constituents of bone, being the same for all tissues discussed herein, we consider the following experiments

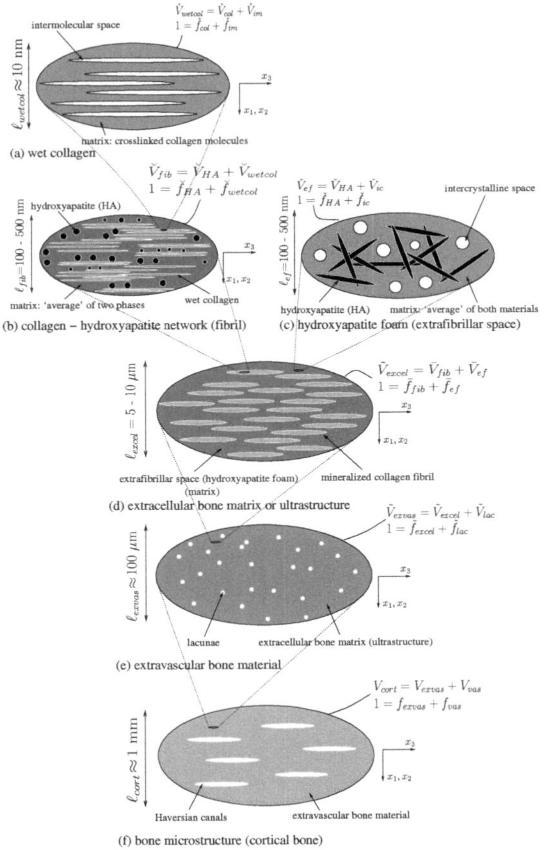

Figure D.2: Micromechanical representation of bone material by means of a six-step homogenization procedure

(see also (Fritsch and Hellmich 2007)): Tests with an ultrasonic interferometer coupled with a solid media pressure apparatus (Katz and Ukraincik 1971; Gilmore and Katz 1982) reveal the isotropic elastic properties of hydroxyapatite powder (Table D.1), which, in view of the largely disordered arrangement of minerals (Lees et al. 1994; Fratzl et al. 1996; Peters et al. 2000; Hellmich and Ulm 2002a), are considered as sufficient for the characterization of the mineral phase (Hellmich and Ulm 2002b; Hellmich et al. 2004b). Given the absence of direct measurements of (molecular) collagen, the elastic properties of (molecular) collagen are approximated by those of dry rat tail tendon, a tissue consisting almost exclusively of collagen. By means of Brillouin light scattering, Cusack and Miller (1979) have determined the respective five independent elastic constants of a transversely isotropic material (Table D.1). We assign the standard bulk modulus of water (Table D.1) to phases comprising water with mechanically insignificant non-collagenous organic matter.

Phase	Bulk modulus k [GPa]	Shear modulus μ [GPa]	Experimental source
Hydroxyapatite	$k_{HA} = 82.6$	$\mu_{HA} = 44.9$	(Katz and Ukraincik 1971)
Water containing non-collagenous organics or osteocytes	$k_{H_2O} = 2.3$	$\mu_{H_2O} = 0$	
	c_{ijkl} [GPa]	c_{ijkl} [GPa]	
Collagen	$c_{col,3333} = 17.9$ $c_{col,1111} = 11.7$	$c_{col,1133} = 7.1$ $c_{col,1122} = 5.1$ $c_{col,1313} = 3.3$	(Cusack and Miller 1979)

Table D.1: 'Universal' (tissue and location-independent) isotropic (or transversely isotropic) stiffness values of elementary constituents

D.4.2 Failure properties of hydroxyapatite crystals and collagen

Recent work on porous hydroxyapatite biomaterials (Fritsch et al. 2009a) has revealed that the elastic limit of single (needle-type) hydroxyapatite crystals can be appropriately characterized through a criterion of the form:

$$\psi = 0, \ldots, 2\pi : \mathfrak{f}_{HA\varphi\vartheta}(\boldsymbol{\sigma}_{HA\varphi\vartheta}) = \beta \max_{\psi} |\sigma_{HA}^{Nn}| + \sigma_{HA}^{NN} - \sigma_{HA}^{ult,t} = 0 \qquad \text{(D.22)}$$

with Euler angles φ and ϑ defining the crystal needle orientation vector $\underline{N}=\underline{e}_r$ in the reference frame ($\underline{e}_1$, $\underline{e}_2$, $\underline{e}_3$), and with ψ defining the orientation of vector $\underline{n}$ related to shear stresses (see Figure D.3). $\beta = \sigma_{HA}^{ult,t}/\sigma_{HA}^{ult,s}$ is the ratio between the uniaxial tensile strength $\sigma_{HA}^{ult,t}$ and the shear strength $\sigma_{HA}^{ult,s}$ of pure hydroxyapatite (abbreviated 'HA'), and $\sigma_{HA}^{Nn} = \underline{N} \cdot \boldsymbol{\sigma}_{HA\varphi\vartheta} \cdot \underline{n}$ and $\sigma_{HA}^{NN} = \underline{N} \cdot \boldsymbol{\sigma}_{HA\varphi\vartheta} \cdot \underline{N}$ are the normal and shear stress components related to a surface with normal $\underline{N}(\varphi, \vartheta)$. These strength values can be gained from experiments of Akao et al.

(1981) and Shareef et al. (1993), see (Fritsch et al. 2009a) for further details, and they amount to 52.2 MPa and 80.3 MPa, respectively (see also Table D.2). Beyond the elastic regime, we consider associated ideal plasticity according to Eq. (D.2) - having in mind a mathematically feasible strategy for mimicking layered water-induced ductile sliding between crystals, which maintains the crystals' stress levels reached at the elastic limit. Use of (D.22) in (D.2) yields the flow and consistency rules as

$$\begin{aligned} \dot{\varepsilon}^p_{HA\varphi\vartheta} &= \dot{\lambda}_{HA}\,[\underline{N}\otimes\underline{N} + \beta\ \mathrm{sgn}(\sigma^{Nn}_{HA})(\underline{N}\otimes\underline{n}+\underline{n}\otimes\underline{N})], \\ \dot{\lambda}_{HA}\left(\beta\max_{\psi}|\sigma^{Nn}_{HA}| + \sigma^{NN}_{HA} - \sigma^{ult,t}_{HA}\right) &= 0, \\ \dot{\lambda}_{HA} &\geq 0, \\ \beta\max_{\psi}|\sigma^{Nn}_{HA}| + \sigma^{NN}_{HA} - \sigma^{ult,t}_{HA} &\leq 0, \end{aligned} \qquad \text{(D.23)}$$

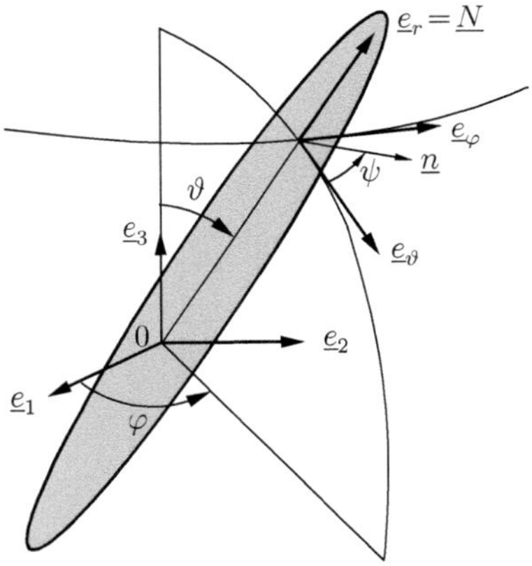

Figure D.3: Cylindrical (needle-like) HA inclusion oriented along vector $\underline{N}$ and inclined by angles ϑ and φ with respect to the reference frame ($\underline{e}_1$, $\underline{e}_2$, $\underline{e}_3$); local base frame ($\underline{e}_r$, $\underline{e}_\vartheta$, $\underline{e}_\varphi$) is attached to the needle.

Phase	Uniaxial tensile strength [MPa]	Uniaxial shear strength [MPa]	Experimental source
Hydroxyapatite	$\sigma^{ult,t}_{HA} = 52.2$	$\sigma^{ult,s}_{HA} = 80.3$	(Akao et al. 1981; Shareef et al. 1993)
Collagen	$\sigma^{ult}_{col} = 144.7$		(Gentleman et al. 2003; Lees et al. 1984a)

Table D.2: 'Universal' (tissue and location-independent) phase strength values

Experiments on collagen fibrils have evidenced the quasi-brittle failure characteristics of this material (Christiansen et al. 2000; Gentleman et al. 2003). Failure of the crosslinks between the cylindrical collagen molecules is standardly agreed upon as the primary cause of collagen failure in the longitudinal direction of the molecules (fibrils) (Buehler 2006; Vesentini et al. 2005). We here represent this fact by a failure criterion of the form

$$\mathfrak{f}_{col}(\boldsymbol{\sigma}_{col}) = |\underline{e}_3 \cdot \boldsymbol{\sigma}_{col} \cdot \underline{e}_3| - \sigma_{col}^{ult} \leq 0 \tag{D.24}$$

where the direction three coincides with the principal orientation direction of collagen (see Figure D.2). Once the equal sign holds in criterion (D.24), we consider that the strengths of both the collagenous phase and of the overall bone materials are reached, while any potential plastic or, more probably, microcracking and crack bridging events leading to toughening in the post-peak regime (Nalla et al. 2004), are beyond the scope of the present manuscript.Given the aforementioned role of the collagen crosslinks for the strength of molecular collagen, a non-mineralized collagenous tissue with crosslinking characteristics close to that of bone is the favorable access to the strength of molecular collagen. As before, we will rely on rat tail tendon, which, under wet conditions, exhibits a strength of 106.1 MPa (Table 2 in (Gentleman et al. 2003)). Again, we have to consider close packing of collagen as to get access to properties of molecular collagen. It is known from neutron diffraction studies (Lees et al. 1984a; Lees 1987a) that diffractional spacing (a measure for the lateral distance of collagen molecules) reduces from 1.5 nm (for wet collagen) to 1.1 nm (for maximally packed (dry) collagen). Accordingly, the cross sectional area of a tensile specimen would reduce by the ratio 1.5/1.1, so that the strength of molecular collagen follows to be 1.5/1.1 times higher than that of wet collagen, i.e. 144.7 MPa (see Table D.2).

D.4.3 Homogenization over wet collagen

An RVE of wet collagen [see Figure D.2(a)] hosts cylindrical intermolecular pores (labeled by suffix '*im*') being embedded into a matrix of crosslinked molecular collagen (labeled by suffix '*col*'), which is suitably considered through a Mori-Tanaka scheme. Unless collagen rupture criterion (D.24) is fulfilled, the RVE behaves purely elastically ($\boldsymbol{\varepsilon}^p_{col}=\boldsymbol{\varepsilon}^p_{im}=\mathbf{0}$), with a homogenized stiffness $\mathbb{C}^{MT}_{wetcol}$ following from specification of (D.6) for r=[col, im]. Thereby, the volume fractions fulfill $\mathring{f}_{im} + \mathring{f}_{col} = 1$, and the concentration tensors $\mathbb{A}_{col}$ and $\mathbb{A}_{im}$, respectively, are given through specification of (D.18) for $\mathbb{P}^0_{im}$=$\mathbb{P}^{col}_{cyl}$, $\mathbb{c}_M$=$\mathbb{c}_{col}$, as well as for $\mathbb{c}_r$= $\mathbb{c}_{col}$ and $\mathbb{c}_r = \mathbb{c}_{im} = 3k_{H_2O}\mathbb{J}$, respectively. $J_{ijkl} = 1/3\delta_{ij}\delta_{kl}$ is the volumetric part of the fourth order unity tensor $\mathbb{I}$; see Table D.1 for k_{H_2O}. According to the aforementioned specifications, the concentration relation (D.17) for the matrix of molecular collagen within an RVE of wet collagen reads as

$$\boldsymbol{\varepsilon}_{col} = \left\{(1 - \mathring{f}_{im})\mathbb{I} + \mathring{f}_{im}\left[\mathbb{I} + \mathbb{P}^{col}_{cyl} : (\mathbb{c}_{im} - \mathbb{c}_{col})\right]^{-1}\right\}^{-1} : \boldsymbol{E}_{wetcol} \tag{D.25}$$

whereby the components of morphology tensor $\mathbb{P}^{col}_{cyl}$ are given in the Appendix.

D.4.4 Homogenization over mineralized collagen fibril

An RVE of mineralized collagen fibrils [see Figure D.2(b)] hosts crystal clusters (represented through spherical hydroxyapatite inclusions, labeled by suffix 'HA') and cylindrical microfibrils of wet collagen (labeled by suffix '$wetcol$'), which are mutually intertwingled. In order to consider this morphology, a self-consistent scheme is appropriate. Unless the wet collagen phase does not fail [see Subsections D.4.3 and D.4.2, in particular Eq. (D.24)], the RVE behaves purely elastically ($\boldsymbol{\varepsilon}^p_{HA}=\boldsymbol{\varepsilon}^p_{wetcol}=\mathbf{0}$), with a homogenized stiffness $\mathbb{C}^{SCS}_{fib}$ following from specification of (D.6), for r=[HA, $wetcol$]. Thereby, the volume fractions fulfill $\breve{f}_{wetcol}+\breve{f}_{HA}=1$, and the concentration tensors $\mathbb{A}_{HA}$ and $\mathbb{A}_{wetcol}$, respectively, are given through specification of (D.13) for $\mathbb{C}^{hom}=\mathbb{C}^{SCS}_{fib}$, for $\mathbb{P}^0_{HA}=\mathbb{P}^{fib}_{sph}$ and $\mathbb{P}^0_{wetcol}=\mathbb{P}^{fib}_{cyl}$, respectively, as well as for $\mathbb{c}_r=\mathbb{c}_{HA}=3k_{HA}\mathbb{J}+2\mu_{HA}\mathbb{K}$, and $\mathbb{c}_r=\mathbb{C}^{MT}_{wetcol}$, respectively. $\mathbb{K}=\mathbb{I}-\mathbb{J}$ is the deviatoric part of the fourth order unity tensor $\mathbb{I}$; see Table D.1 for k_{HA} and μ_{HA}. According to the aforementioned specifications, the concentration relation (D.12) for the phase 'wet collagen' within an RVE of mineralized collagen fibril reads as

$$\begin{aligned}\boldsymbol{\varepsilon}_{wetcol} = &\left[\mathbb{I}+\mathbb{P}^{fib}_{cyl}:\left(\mathbb{C}^{MT}_{wetcol}-\mathbb{C}^{SCS}_{fib}\right)\right]^{-1}: \\ &\left\{\breve{f}_{wetcol}\left[\mathbb{I}+\mathbb{P}^{fib}_{cyl}:\left(\mathbb{C}^{MT}_{wetcol}-\mathbb{C}^{SCS}_{fib}\right)\right]^{-1}+\right. \\ &\left.\breve{f}_{HA}\left[\mathbb{I}+\mathbb{P}^{fib}_{sph}:\left(\mathbb{c}_{HA}-\mathbb{C}^{SCS}_{fib}\right)\right]^{-1}\right\}^{-1}:\boldsymbol{E}_{fib}\end{aligned} \tag{D.26}$$

whereby the components of $\mathbb{P}^{fib}_{sph}$ and $\mathbb{P}^{fib}_{cyl}$ are given in the Appendix – and $\boldsymbol{\varepsilon}_{wetcol}$ (here the 'microscopic' strain) is identical to $\boldsymbol{E}_{wetcol}$ of Eq. (D.25), there being the 'macroscopic' strain.

D.4.5 Homogenization over extrafibrillar space (hydroxyapatite foam)

An RVE of extrafibrillar space [see Figure D.2(c)] hosts crystal needles (represented through cylindrical hydroxyapatite inclusions, labeled by suffix 'HA') being oriented in all space directions, and spherical, water-filled pores (intercrystalline space, labeled by suffix 'ic'). The corresponding polycrystal-type morphology is appropriately represented through a self-consistent scheme. Sliding between crystals is modeled through criterion (D.23), leading to plastic strains $\boldsymbol{\varepsilon}^p_{HA}$, and no plasticity occurs in the intercrystalline space ($\boldsymbol{\varepsilon}^p_{ic}=\mathbf{0}$). The homogenized stiffness of an RVE of extrafibrillar space, $\mathbb{C}^{SCSII}_{ef}$, follows from specification of (D.6) for r=[HA, ic]. Thereby, the volume fractions fulfill $\breve{f}_{HA}+\breve{f}_{ic}=1$, and the concentration tensors $\mathbb{A}_{HA\varphi\vartheta}$ and $\mathbb{A}_{ic}$, respectively, are given through specification of (D.13) for $\mathbb{C}^{hom}=\mathbb{C}^{SCSII}_{ef}$, for $\mathbb{P}^0_{HA}=\mathbb{P}^{ef}_{cyl}(\vartheta,\varphi)$ and $\mathbb{P}^0_{ic}=\mathbb{P}^{ef}_{sph}$, respectively, as well as for $\mathbb{c}_r=\mathbb{c}_{HA}$ and $\mathbb{c}_r=\mathbb{c}_{ic}=3k_{H_2O}\mathbb{J}$ (see Table D.1), respectively. Thereby, summation over all crystal orientations is done by integration over Euler angles $\vartheta=0,\ldots,\pi$ and $\varphi=0,\ldots,2\pi$. Accordingly, the concentration-influence relation (D.17) for the hydroxyapatite phase oriented in a specific direction (ϑ, φ) within an RVE of extrafibrillar space reads as

$$\varepsilon_{HA\varphi\vartheta} = [\mathbb{I} + \mathbb{P}^{ef}_{cyl}(\vartheta,\varphi) : (\mathbb{c}_{HA} - \mathbb{C}^{SCSII}_{ef})]^{-1} :$$
$$\left\{\left\{\check{f}_{HA} \int_{\phi=0}^{2\pi}\int_{\theta=0}^{\pi} [\mathbb{I} + \mathbb{P}^{ef}_{cyl}(\theta,\phi) : (\mathbb{c}_{HA} - \mathbb{C}^{SCSII}_{ef})]^{-1} \frac{\sin\theta \,\mathrm{d}\theta \,\mathrm{d}\phi}{4\pi} + \right.\right.$$
$$\left. + \check{f}_{ic}[\mathbb{I} + \mathbb{P}^{ef}_{sph} : (\mathbb{c}_{ic} - \mathbb{C}^{SCSII}_{ef})]^{-1}\right\}^{-1} :$$
$$\left\{ \boldsymbol{E}_{ef} - \check{f}_{HA} \int_{\phi=0}^{2\pi}\int_{\theta=0}^{\pi} [\mathbb{I} + \mathbb{P}^{ef}_{cyl}(\theta,\phi) : (\mathbb{c}_{HA} - \mathbb{C}^{SCSII}_{ef})]^{-1} : \right.$$
$$\left. \mathbb{P}^{ef}_{cyl}(\theta,\phi) : \mathbb{c}_{HA} : \boldsymbol{\varepsilon}^{p}_{HA\vartheta\varphi} \frac{\sin\theta \,\mathrm{d}\theta \,\mathrm{d}\phi}{4\pi} \right\} + \mathbb{P}^{ef}_{cyl}(\vartheta,\varphi) : \mathbb{c}_{HA} : \boldsymbol{\varepsilon}^{p}_{HA\vartheta\varphi}\Big\} \qquad \text{(D.27)}$$

whereby the components of $\mathbb{P}^{ef}_{sph}$ and $\mathbb{P}^{ef}_{cyl}$ are given in the Appendix. According to (D.8) applied to the present homogenization step, plastic strains $\boldsymbol{\varepsilon}^{p}_{HA}$ in the hydroxyapatite phases imply a plastic strain $\boldsymbol{E}^{p}_{ef}$ at the level of the RVE of extrafibrillar space.

D.4.6 Homogenization over extracellular bone matrix

An RVE of extracellular bone matrix or ultrastructure [see Figure D.2(d)] hosts cylindrical mineralized fibrils (labeled by suffix 'fib') being embedded into a matrix of extrafibrillar space (labeled by suffix 'ef'). This morphology is suitably modeled by means of a Mori-Tanaka scheme. As discussed in the previous Subsection D.4.5, the extrafibrillar matrix may be subjected to plastic strains, while we do not consider plastic strains in the mineralized fibrils ($\boldsymbol{\varepsilon}^{p}_{fib} = \mathbf{0}$). The homogenized stiffness of an RVE of extracellular bone matrix, $\mathbb{C}^{MTII}_{excel}$, follows from specification of (D.6) for r=[fib, ef]. Thereby, the volume fractions fulfill $\bar{f}_{fib} + \bar{f}_{ef} = 1$, and the concentration tensors $\mathbb{A}_{fib}$ and $\mathbb{A}_{ef}$, respectively, follow from specification of (D.18) for $\mathbb{c}_M = \mathbb{C}^{SCSII}_{ef}$, for $\mathbb{P}^0_{fib} = \mathbb{P}^{ef}_{cyl}$, as well as for $\mathbb{c}_r = \mathbb{C}^{SCS}_{fib}$ and $\mathbb{c}_r = \mathbb{C}^{SCSII}_{ef}$, respectively. Accordingly, the concentration influence relation (D.17) for the phase extrafibrillar space within an RVE of extracellular bone matrix reads as

$$\varepsilon_{ef} = \left\{\bar{f}_{ef}\mathbb{I} + \bar{f}_{fib}[\mathbb{I} + \mathbb{P}^{ef}_{cyl} : (\mathbb{C}^{SCS}_{fib} - \mathbb{C}^{SCSII}_{ef})]^{-1}\right\}^{-1} :$$
$$\left\{\boldsymbol{E}_{excel} - \bar{f}_{fib}[\mathbb{I} + \mathbb{P}^{ef}_{cyl} : (\mathbb{C}^{SCS}_{fib} - \mathbb{C}^{SCSII}_{ef})]^{-1} : \mathbb{P}^{ef}_{cyl} : (-\mathbb{C}^{SCSII}_{ef} : \boldsymbol{\varepsilon}^{p}_{ef})\right\} \qquad \text{(D.28)}$$

whereby the components of $\mathbb{P}^{ef}_{cyl}$ are given in the Appendix. According to (D.8) applied to the present homogenization step, plastic strains in the extrafibrillar space (see Subsection D.4.5, $\boldsymbol{\varepsilon}^{p}_{ef} = \boldsymbol{E}^{p}_{ef}$) imply a plastic strain $\boldsymbol{E}^{p}_{excel}$ at the level of the RVE of the extracellular bone matrix.

D.4.7 Homogenization over extravascular bone material

An RVE of extravascular bone material [see Figure D.2(e)] hosts spherical empty pores called lacunae (labeled by suffix '*lac*') being embedded into a matrix of extracellular bone matrix (labeled by suffix '*excel*'). This morphology is suitably modeled by means of a Mori-Tanaka scheme. As discussed in the previous Subsection D.4.6, the extracellular bone matrix may be subjected to plastic strains while we do not consider plastic strains in the lacunae ($\boldsymbol{\varepsilon}^p_{lac} = \mathbf{0}$). The homogenized stiffness of an RVE of extravascular bone material, $\mathbb{C}^{MTIII}_{exvas}$, follows from specification of (D.6) for r=[lac, $excel$]. Thereby, the volume fractions fulfill $\tilde{f}_{lac} + \tilde{f}_{excel} = 1$, and the concentration tensors $\mathbb{A}_{lac}$ and $\mathbb{A}_{excel}$, respectively, follow from specification of (D.18) for $\mathbb{c}_M$=$\mathbb{C}^{MTII}_{excel}$, for $\mathbb{P}^0_{lac}$=$\mathbb{P}^{excel}_{sph}$, as well as for $\mathbb{c}_r$=$\mathbb{c}_{lac}$=$\mathbb{0}$ and $\mathbb{c}_r$=$\mathbb{C}^{MTII}_{excel}$, respectively. $\mathbb{c}_{lac} = \mathbb{0}$ relates to the fact that the lacunar pores are empty (drained) in all experiments considered in Section D.6 – for undrained situations, $\mathbb{c}_{lac} = 3k_{H_2O}\mathbb{J}$ would be appropriate, see (Fritsch and Hellmich 2007) for details. According to the aforementioned specifications, the concentration-influence relation (D.17) for the phase 'extrafibrillar space' within an RVE of extracellular bone matrix reads as

$$\boldsymbol{\varepsilon}_{excel} = \left\{\tilde{f}_{excel}\mathbb{I} + \tilde{f}_{lac}[\mathbb{I} - \mathbb{P}^{excel}_{sph} : \mathbb{C}^{MTII}_{excel}]^{-1}\right\}^{-1} :$$

$$\left\{\boldsymbol{E}_{exvas} - \tilde{f}_{lac}[\mathbb{I} - \mathbb{P}^{excel}_{sph} : \mathbb{C}^{MTII}_{excel}]^{-1} : \mathbb{P}^{excel}_{sph} : (-\mathbb{C}^{MTII}_{excel} : \boldsymbol{\varepsilon}^p_{excel})\right\} \tag{D.29}$$

whereby the components of $\mathbb{P}^{excel}_{sph}$ are given in the Appendix. According to (D.8) applied to the present homogenization step, plastic strains in the extracellular bone matrix (see Subsection D.4.6, $\boldsymbol{\varepsilon}^p_{excel}$=$\boldsymbol{E}^p_{excel}$) imply a plastic strain $\boldsymbol{E}^p_{exvas}$ at the level of the RVE of the extravascular bone material.

D.4.8 Homogenization over cortical bone material

An RVE of cortical bone material [see Figure D.2(f)] hosts cylindrical empty pores called Haversian canals or vascular space (labeled by suffix '*vas*') being embedded into a matrix of extravascular bone material (labeled by suffix '*exvas*'). This morphology is suitably modeled by means of a Mori-Tanaka scheme. As discussed in the previous Subsection D.4.7, the extravascular bone material may be subjected to plastic strains, while we do not consider plastic strains in the Haversian canals ($\boldsymbol{\varepsilon}^p_{vas} = \mathbf{0}$). The homogenized stiffness of an RVE of cortical bone material, $\mathbb{C}^{MTIV}_{cort}$, follows from specification of (D.6) for r=[vas, $exvas$]. Thereby, the volume fractions fulfill $f_{vas} + f_{exvas} = 1$, and the concentration tensors $\mathbb{A}_{vas}$ and $\mathbb{A}_{exvas}$, respectively, follow from specification of (D.18) for $\mathbb{c}_M$=$\mathbb{C}^{MTIII}_{exvas}$, for $\mathbb{P}^0_{vas}$=$\mathbb{P}^{exvas}_{cyl}$, as well as for $\mathbb{c}_r$=$\mathbb{c}_{vas}$=$\mathbb{0}$ and $\mathbb{c}_r$=$\mathbb{C}^{MTIII}_{exvas}$, respectively. $\mathbb{c}_{vas} = \mathbb{0}$ relates to the fact that the Haversian canals are empty (drained) in all experiments considered in Section D.6. According to the aforementioned specifications, the concentration-influence relation (D.17) for the phase 'extravascular bone material' within an RVE of cortical bone material reads as

$$\varepsilon_{exvas} = \left\{ f_{exvas}\mathbb{I} + f_{vas}[\mathbb{I} - \mathbb{P}^{exvas}_{cyl} : \mathbb{C}^{MTIII}_{exvas}]^{-1} \right\}^{-1} :$$

$$\left\{ \boldsymbol{E}_{cort} - f_{vas}[\mathbb{I} - \mathbb{P}^{exvas}_{cyl} : \mathbb{C}^{MTIII}_{exvas}]^{-1} : \mathbb{P}^{exvas}_{cyl} : (-\mathbb{C}^{MTIII}_{exvas} : \varepsilon^{p}_{exvas}) \right\} \tag{D.30}$$

whereby the components of $\mathbb{P}^{exvas}_{cyl}$ are given in the Appendix. According to (D.8) applied to the present homogenization step, plastic strains in the extravascular bone material (see Subsection D.4.7, $\varepsilon^p_{exvas}=\boldsymbol{E}^p_{exvas}$) imply a plastic strain $\boldsymbol{E}^p_{cort}$ at the level of the RVE of the cortical bone material.

D.5 Algorithmic aspects

We are left with using the partially incremental constitutive relations developed in Sections D.3 and D.4 for computation of stress-strain relations. This requires some algorithmic deliberations which we will describe in view of a stress-strain curve for uniaxial stress applied to an RVE of cortical bone, $\boldsymbol{\Sigma}_{cort} = \Sigma_{33}\underline{e}_3 \otimes \underline{e}_3$, the loading direction $\underline{e}_3$ coinciding with the longitudinal (axial) direction of the bone material (see Figure D.2). This stress is applied in load increments labeled by n, starting at $\Sigma_{33} = 0$, and being accumulated up to failure of the material. Accordingly, flow rule (D.2) and (D.23) is considered in a discretized fashion: It is evaluated for a finite number of needle orientation directions ('families'), and it is integrated over the n-th load step,

$$\Delta\varepsilon^p_{HA\varphi\vartheta,n+1} = \Delta\lambda_{HA,n+1}[\underline{N} \otimes \underline{N} + \beta \,\mathrm{sgn}(\sigma^{Nn}_{HA})(\underline{N} \otimes \underline{n} + \underline{n} \otimes \underline{N})] \tag{D.31}$$

with

$$\varepsilon^p_{HA\varphi\vartheta,n+1} = \varepsilon^p_{HA\varphi\vartheta,n} + \Delta\varepsilon^p_{HA\varphi\vartheta,n+1} \tag{D.32}$$

At the beginning of the very first load step, there are neither plastic strains ($\boldsymbol{E}^p_{cort,0} = \mathbf{0}$) nor total strains ($\boldsymbol{E}_{cort,0} = \mathbf{0}$); at the end of an arbitrary later load step with label n, there may be plastic strains $\boldsymbol{E}^p_{cort,n}$ and total strains $\boldsymbol{E}_{cort,n}$, both related to stresses $\boldsymbol{\Sigma}_{cort,n} = \Sigma_{33,n}\underline{e}_3 \otimes \underline{e}_3$. Then, the general task is to compute the strain increments $\Delta\boldsymbol{E}^p_{cort,n+1}$ and $\Delta\boldsymbol{E}_{cort,n+1}$, leading to total strains $\boldsymbol{E}^p_{cort,n+1} = \boldsymbol{E}^p_{cort,n} + \Delta\boldsymbol{E}^p_{cort,n+1}$ and $\boldsymbol{E}_{cort,n+1} = \boldsymbol{E}_{cort,n} + \Delta\boldsymbol{E}_{cort,n+1}$, following from the stress increment $\Delta\boldsymbol{\Sigma}_{cort,n+1} = \Delta\Sigma_{33,n+1}\underline{e}_3 \otimes \underline{e}_3$.

To fulfill this task, an iterative procedure is applied: First, the macroscopic strains are estimated from specification of (D.7) for an RVE of cortical bone, on the assumption that no plastic strains would occur during the $(n+1)$-st load step, which may be referred to as a 'trial step' in the line of classical computational elastoplasticity (Simo and Taylor 1985),

$$\boldsymbol{E}^{trial}_{cort,n+1} = \mathbb{C}^{MTIV}_{cort} : \boldsymbol{\Sigma}_{cort,n+1} + \boldsymbol{E}^p_{cort,n} \tag{D.33}$$

Then, these trial strains are concentrated into the lower-scale RVEs, by means of Eqs. (D.25)-(D.30), all specified for $\boldsymbol{E}_{cort} = \boldsymbol{E}^{trial}_{cort,n+1}$; $\varepsilon^p_{exvas} = \varepsilon^p_{exvas,n}$, $\varepsilon_{exvas} = \boldsymbol{E}_{exvas} = \varepsilon^{trial}_{exvas,n+1} =$

$\boldsymbol{E}^{trial}_{exvas,n+1}$; $\boldsymbol{\varepsilon}^{p}_{excel} = \boldsymbol{\varepsilon}^{p}_{excel,n}$, $\boldsymbol{\varepsilon}_{excel} = \boldsymbol{E}_{excel} = \boldsymbol{\varepsilon}^{trial}_{excel,n+1} = \boldsymbol{E}^{trial}_{excel,n+1}$; $\boldsymbol{\varepsilon}^{p}_{ef} = \boldsymbol{\varepsilon}^{p}_{ef,n}$, $\boldsymbol{\varepsilon}_{ef} = \boldsymbol{E}_{ef} = \boldsymbol{\varepsilon}^{trial}_{ef,n+1} = \boldsymbol{E}^{trial}_{ef,n+1}$; $\boldsymbol{\varepsilon}^{p}_{HA\varphi\vartheta} = \boldsymbol{\varepsilon}^{p}_{HA\varphi\vartheta,n}$, $\boldsymbol{\varepsilon}_{HA\varphi\vartheta} = \boldsymbol{\varepsilon}^{trial}_{HA\varphi\vartheta,n+1}$. Within the RVE of extrafibrillar material, the trial stress states in hydroxyapatite phases follow to be

$$\boldsymbol{\sigma}^{trial}_{HA\varphi\vartheta,n+1} = \mathbb{c}_{HA} : [\boldsymbol{\varepsilon}^{trial}_{HA\varphi\vartheta,n+1} - \boldsymbol{\varepsilon}^{p}_{HA\varphi\vartheta,n}] \tag{D.34}$$

and this trial stress allows one to identify the plasticizing mineral phases in load step $n+1$:

$$\begin{aligned} \mathfrak{f}_{HA\varphi\vartheta}(\boldsymbol{\sigma}^{trial}_{HA\varphi\vartheta,n+1}) \leq 0 &\leftrightarrow \Delta\lambda_{HA\varphi\vartheta,n+1} = 0 \\ \mathfrak{f}_{HA\varphi\vartheta}(\boldsymbol{\sigma}^{trial}_{HA\varphi\vartheta,n+1}) > 0 &\leftrightarrow \Delta\lambda_{HA\varphi\vartheta,n+1} > 0 \end{aligned} \tag{D.35}$$

In the first case, the load step is elastic, $\Delta\boldsymbol{E}^{p}_{cort,n+1} = \mathbf{0}$ and $\boldsymbol{E}^{trial}_{cort,n+1} = \boldsymbol{E}_{cort,n+1}$, and the computation can proceed to the next load step, $n+2$. In the second case, the load step is elastoplastic, the plastic multiplier $\Delta\lambda_{HA\varphi\vartheta,n+1}$ and the plastic strain increment $\Delta\boldsymbol{\varepsilon}^{p}_{HA\varphi\vartheta,n+1}$ need to be determined. In the line of classical computational elastoplasticity, this is done by means of the so-called return map algorithm, also called projection algorithm (Simo and Taylor 1985): A trial stress state $\boldsymbol{\sigma}^{trial}_{HA\varphi\vartheta,n+1}$ which lies outside the elastic domain has to be projected back onto the failure surface $\mathfrak{f}_{HA\varphi\vartheta} = \mathfrak{f}_1$ in Fig. D.4, which gives a first approximation of the stresses in the HA phase,

$$\begin{aligned} &\boldsymbol{\sigma}^{(1)}_{HA\varphi\vartheta,n+1} = \boldsymbol{\sigma}^{trial}_{HA\varphi\vartheta,n+1} - \mathbb{c}_{HA} : \Delta\lambda_{HA\varphi\vartheta,n+1}[\underline{N} \otimes \underline{N} + \\ &+\beta\,\mathrm{sgn}(\sigma^{Nn}_{HA})(\underline{N} \otimes \underline{n} + \underline{n} \otimes \underline{N})], \\ &\mathfrak{f}(\boldsymbol{\sigma}^{(1)}_{HA,n+1}) = 0 \\ &\rightarrow \Delta\lambda_{HA\varphi\vartheta,n+1} = \\ &= \frac{(3k_{HA} - 2\mu_{HA})\bar{\varepsilon}_{11} + (3k_{HA} - 2\mu_{HA})\bar{\varepsilon}_{22} + (3k_{HA} + 4\mu_{HA})\bar{\varepsilon}_{33}}{3k_{HA} + 4\mu_{HA} + 6\beta^2\mu} + \\ &+\frac{\mathrm{sgn}(\sigma^{Nn}_{HA})6\beta\,\mu\,\bar{\varepsilon}_{13} - 3\sigma^{ult,t}_{HA}}{3k_{HA} + 4\mu_{HA} + 6\beta^2\mu} \end{aligned} \tag{D.36}$$

whereby the components of the difference $(\boldsymbol{\varepsilon}_{HA\varphi\vartheta,n+1} - \boldsymbol{\varepsilon}^{p}_{HA\varphi\vartheta,n})$, $\bar{\varepsilon}_{ij}$, are given in a local base frame $(\underline{e}_r, \underline{e}_\vartheta, \underline{e}_\varphi)$, see Fig. D.3.

Use of $\Delta\lambda_{HA\varphi\vartheta,n+1}$ in (D.31), and insertion of the result into (D.8) specified for the extrafibrillar RVE, for the extracellular RVE, for the extravascular RVE, and for the cortical RVE, yields a first approximation of $\boldsymbol{E}^{p(1)}_{cort,n+1}$ and $\Delta\boldsymbol{E}^{p(1)}_{cort,n+1}$. These plastic strains are inserted into (D.33) where $\boldsymbol{E}^{p}_{cort,n}$ is replaced by $\boldsymbol{E}^{p(1)}_{cort,n+1}$, and the aforementioned procedure is repeated, leading to strains $\Delta\boldsymbol{E}^{p(2)}_{cort,n+1}$ and $\boldsymbol{E}^{p(2)}_{cort,n+1}$. Further repetitions of the aforementioned procedure are performed, the k-th performance yielding strains $\boldsymbol{E}^{p(k)}_{cort,n+1}$; and this is done until $\Delta\boldsymbol{E}^{p(k)}_{cort,n+1}$ approaches zero up to a prescribed tolerance so that satisfactorily precise values for $\boldsymbol{E}^{p}_{cort,n+1}$ and $\boldsymbol{E}_{cort,n+1}$ have been attained. Then, the next load step, $(n+2)$, is tackled.

A particular case deserves further discussion: If the trial stress state $\boldsymbol{\sigma}^{trial}_{HA\varphi\vartheta,n+1}$ lies within the gray shaded area of Fig. D.4, projection step (D.36) may deliver negative values for $|\sigma_{Nn}|$, which is not admissible. In this case, a two-surface failure criterion is employed, the second surface being defined through

$$\mathfrak{f}_{2,HA\varphi\vartheta}(\boldsymbol{\sigma}_{HA\varphi\vartheta,n+1}) = \sigma^{NN}_{HA} - \sigma^{ult,t}_{NN,HA} = 0, \tag{D.37}$$

and Eq. (D.31) is extended according to Koiter's flow rule (Koiter 1960)

$$\Delta\boldsymbol{\varepsilon}^{p}_{HA\varphi\vartheta,n+1} = \Delta\lambda_{1,HA\varphi\vartheta,n+1}\frac{\partial\mathfrak{f}_{1,HA\varphi\vartheta}}{\partial\boldsymbol{\sigma}_{HA\varphi\vartheta,n+1}} + \Delta\lambda_{2,HA\varphi\vartheta,n+1}\frac{\partial\mathfrak{f}_{2,HA\varphi\vartheta}}{\partial\boldsymbol{\sigma}_{HA\varphi\vartheta,n+1}} \tag{D.38}$$

with $\mathfrak{f}_{1,HA\varphi\vartheta} = \mathfrak{f}_{HA\varphi\vartheta} = 0$ from Eq. (D.22). This leads to plastic multipliers reading as

$$\begin{aligned}
&\Delta\lambda_{1,HA\varphi\vartheta,n+1} = \text{sgn}(\sigma^{Nn}_{HA})\frac{\bar{\varepsilon}_{13}}{\beta}\\
&\Delta\lambda_{2,HA\varphi\vartheta,n+1} =\\
&\frac{(3k_{HA}-2\mu_{HA})\beta\bar{\varepsilon}_{11} + (3k_{HA}-2\mu_{HA})\beta\bar{\varepsilon}_{22} + (3k_{HA}+4\mu_{HA})\beta\bar{\varepsilon}_{33}+}{(3k_{HA}+4\mu_{HA})\beta} +\\
&+\frac{\text{sgn}(\sigma^{Nn}_{HA})(3k_{HA}+4\mu_{HA})\bar{\varepsilon}_{13} - 3\beta\sigma^{ult,t}_{HA}}{(3k_{HA}+4\mu_{HA})\beta}
\end{aligned} \tag{D.39}$$

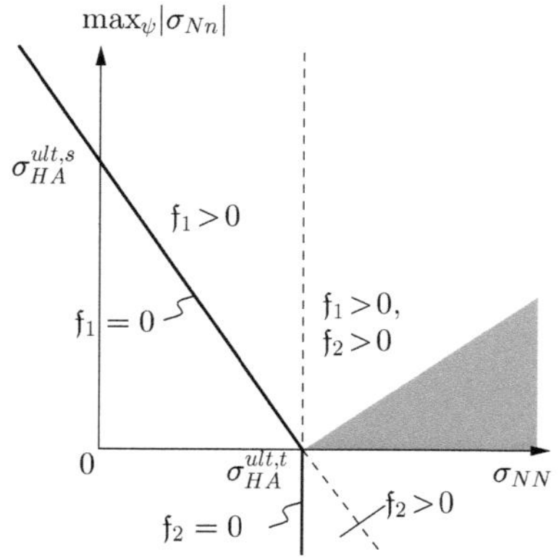

Figure D.4: Schematic representation of the loading surfaces $\mathfrak{f}_1 = \mathfrak{f}_{1,HA\varphi\vartheta}$ and $\mathfrak{f}_2 = \mathfrak{f}_{2,HA\varphi\vartheta}$, for a specific needle family with orientation given through φ and ϑ, in the σ_{NN}-σ_{Nn} stress space.

D.6 Experimental validation of multiscale model for bone strength

The mathematical model developed in Sections D.4 and D.5 is based on experimentally determined elasticity and strength properties of the elementary components hydroxyapatite, (molecular) collagen, and water. This model predicts, for each set of tissue-specific volume fractions $\mathring{f}_{col}$, $\check{f}_{wetcol}$, $\check{f}_{HA}$, $\bar{f}_{fib}$, $\tilde{f}_{excel}$, and f_{exvas} (see Figure D.2), the corresponding tissue-specific elasticity and strength properties at all observation scales of Figure D.2. Thus, a strict experimental validation of the mathematical model is realized as follows: (i) different sets of volume fractions are determined from composition experiments on different bone samples with different ages, from different species and different anatomical locations (micrographs, weighing tests on demineralized/dehydrated tissues, neutron diffraction tests; see Subsection D.6.1); (ii) these volume fractions are used as model input, and (iii) corresponding model-predicted strength values (model output) are compared to results from strength experiments on the same or very similar bone samples. We here refrain from validation of model-predicted elastic values, since these are reported, in great detail, in (Fritsch and Hellmich 2007).

D.6.1 Experimental set providing tissue-specific volume fractions as model input

Experimental validation of the six-step upscaling procedure [Eqs. (D.22) to (D.39)] requires determination of the phase volume fractions within the six considered RVEs (Figure D.2).

Within an RVE of cortical bone [Figure D.2(f)], the extravascular volume fraction f_{exvas} is primarily driven by the interplay of osteoblastic and osteoclastic action in the vascular pore space. We here have access to typical mammalian cortical bone under physiological conditions, for which f_{exvas} does not exceed 5% (Sietsema 1995), and the microradiographs of bovine tibia provided by Lees et al. (1979a) yield f_{exvas}=3% (see (Fritsch and Hellmich 2007) for details); we will adopt this value throughout this validation section.

Within an RVE of extravascular bone material [Figure D.2(e)], the lacunar volume fraction $\tilde{f}_{lac}$ relates to the way osteoblasts work: when laying down osteoid, a typical fraction of osteoblasts become buried in this newly formed ultrastructure, leading to the formation of lacunae. Hence, $\tilde{f}_{lac}$ always lies in a narrow range of values, around $\tilde{f}_{lac}$=2% (see (Fritsch and Hellmich 2007) for details); we will adopt this value for the remainder of this validation section.

As regards hydroxyapatite and collagen contents, Lees (1987b) has provided the weight fractions of mineral and organic components within cortical bone samples, WF_{HA}^{cort} and WF_{org}^{cort}, for several mammalian species and organs, including human and bovine bone samples, together with their mass densities ρ_{cort} (see Table D.3). These values give access to the weight fractions at the extracellular (ultrastructural) scale [Figure D.2(d)], through

tissue	ρ_{cort} [g/cm^3] given	WF_{HA}^{cort} [-] given	WF_{org}^{cort} [-] given	$\bar{f}_{HA}$ [-] Eqs. (39), (42), (43)	$\bar{f}_{col}$ [-] Eqs. (40)-(42), (44)
human femur	1.98[a]	0.655[a]	0.227[a]	0.46	0.30
human tibia	1.98[a]	0.659[a]	0.228[a]	0.46	0.30
bovine femur	2.105[a]	0.717[a]	0.180[a]	0.53	0.25
bovine tibia	2.02[a]	0.667[a]	0.209[a]	0.47	0.28
equine radius	2.015[b]	-	-	0.47[c]	0.27[c]

tissue	d_s [nm] Eqs. (42), (49)	$\bar{f}_{fib}$ [-] Eq. (48)	f_{HA} [-] Eqs. (50), (52)	$\check{f}_{HA}$ [-] Eqs. (50), (51)	$\mathring{f}_{col}$ [-] Eqs. (53), (54)
human femur	1.25	0.53	0.65	0.28	0.42
human tibia	1.25	0.53	0.66	0.28	0.42
bovine femur	1.23	0.44	0.71	0.30	0.36
bovine tibia	1.24	0.49	0.66	0.28	0.39
equine radius	1.25	0.48	0.65	0.28	0.38

[a] experimental data: (Lees 1987b), Table 2

[b] experimental data: (Riggs et al. 1993)

[c] calculated with Eqs. (45)-(47)

Table D.3: Tissue-specific composition values

$$WF_{HA}^{excel} = \frac{WF_{HA}^{cort}}{1 - \frac{\rho_{H_2O} \times [f_{vas} + (1 - f_{vas})\tilde{f}_{lac}]}{\rho_{cort}}} \tag{D.40}$$

$$WF_{org}^{excel} = \frac{WF_{org}^{cort}}{1 - \frac{\rho_{H_2O} \times [f_{vas} + (1 - f_{vas})\tilde{f}_{lac}]}{\rho_{cort}}} \tag{D.41}$$

with $\rho_{H_2O} = 1$ kg/dm^3 as the mass density of water filling the vascular and lacunar pores spaces. Since 90% of mass of organic matter in bone is collagen (Urist et al. 1983; Lees 1987a; Weiner and Wagner 1998), the weight fraction of collagen within the extracellular matrix follows to be

$$WF_{col}^{excel} = 0.9 \times WF_{org}^{excel}, \tag{D.42}$$

These weight fractions, together with the tissue mass density at the extracellular scale (the pores of specimens discussed in Table D.3 are filled with water, see (Fritsch and Hellmich 2007) for details),

$$\rho_{excel} = \frac{\rho_{cort} - \rho_{H_2O}[f_{vas} + (1 - f_{vas})\tilde{f}_{lac}]}{1 - f_{vas} - (1 - f_{vas})\tilde{f}_{lac}} \tag{D.43}$$

give access to the mineral and collagen volume fractions at the extracellular observation scale,

$$\bar{f}_{HA} = \frac{\rho_{excel}}{\rho_{HA}} \times WF_{HA}^{excel} \tag{D.44}$$

$$\bar{f}_{col} = \frac{\rho_{excel}}{\rho_{col}} \times WF_{col}^{excel} \tag{D.45}$$

where ρ_{HA}=3.00 kg/dm^3 (Lees 1987a; Hellmich 2004) and ρ_{col} = 1.41 kg/dm^3 (Katz and Li 1973; Lees 1987a) (see Table D.3 for values of $\bar{f}_{HA}$ and $\bar{f}_{col}$ used for the validation of the herein proposed strength model).

The dehydration–demineralization tests of Lees et al. (1979b); Lees (1987a); Lees et al. (1995) show that, throughout samples from the entire vertebrate animal kingdom, the extracellular volume fraction $\bar{f}_{HA}$ depends linearly on the extracellular mass density ρ_{excel},

$$\mathcal{F}_{\bar{f}_{HA}} = \mathcal{A} \times \rho_{excel} + \mathcal{B} \tag{D.46}$$

with $\mathcal{A} = 0.59$ ml/g and $\mathcal{B} = -0.75$, see (Fritsch and Hellmich 2007) for details. Combination of (D.46) with

$$\rho_{excel} = \bar{f}_{H_2O}\,\rho_{H_2O} + \bar{f}_{org}\,\rho_{org} + \bar{f}_{HA}\,\rho_{HA} \tag{D.47}$$

with $\rho_{org} \approx \rho_{col}$, with $1 = \bar{f}_{org} + \bar{f}_{H_2O} + \bar{f}_{HA}$, and with $\bar{f}_{col} = 0.9 \times \bar{f}_{org}$, yields the collagen content as a function of the extracellular mass density,

$$\begin{aligned} \mathcal{F}_{\bar{f}_{col}}(\rho_{excel}) = & \frac{0.9}{\rho_{H_2O} - \rho_{org}} \times \\ & \left\{\mathcal{F}_{\bar{f}_{HA}}(\rho_{excel}) \times [\rho_{HA} - \rho_{H_2O}] - \rho_{excel} + \rho_{H_2O}\right\} \end{aligned} \tag{D.48}$$

see Table D.3 for values based on these functions, used for the validation of the herein proposed strength model.

The extracellular volume fractions of the fibrils and the extracellular space, $\bar{f}_{fib}$ and $\bar{f}_{ef}$ [Figure D.2(d)], can be quantified on the basis of the generalized packing model of Lees et al. (1984b); Lees (1987a), through

$$\bar{f}_{fib} = \bar{f}_{col} \times \frac{v_{fib}}{v_{col}}, \qquad v_{fib} = b\,d_s\,5D \tag{D.49}$$

where $\bar{f}_{col}$ is determined according to (D.45) and (D.42), or according to (D.48) and (D.46), respectively. $v_{col} = 335.6\ nm^3$ is the volume of a single collagen molecule (Lees 1987a). v_{fib} is the volume of one rhomboidal fibrillar unit with length $5D$, width b, and height d_s. b=1.47 nm is an average (rigid) collagen crosslink length valid for all mineralized tissues (Lees et al. 1984b), $D \approx 64$ nm is the axial macroperiod of staggered assemblies of type I collagen, and d_s is the tissue-specific neutron diffraction spacing between collagen molecules, which depends on the mineralization and the hydration state of the tissue (Lees et al. 1984a; Bonar et al. 1985; Lees et al. 1994). For wet tissues, d_s can be given in a dimensionless form (Hellmich and Ulm 2003), as a function of ρ_{excel} only. For the rather narrow range of tissue mass densities considered here, this function can be linearly approximated through

$$d_s = \mathcal{C} \times \rho_{excel} + \mathcal{D} \tag{D.50}$$

where $\mathcal{C} = -0.2000$ nm/(g cm^{-3}) and $\mathcal{D} = 1.6580$ nm.

The volume fractions for scales below the extracellular bone matrix can be derived directly from $\bar{f}_{fib}$ and $\bar{f}_{col}$, on the basis of the finding of Hellmich and Ulm (2001, 2003) that the average hydroxyapatite concentration in the extra-collagenous space of the extracellular bone matrix of wet mineralized tissues is the same inside and outside the fibrils. Accordingly, the relative amount of hydroxyapatite in the extrafibrillar space reads as (Hellmich and Ulm 2001, 2003)

$$\phi_{HA,ef} = \frac{1 - \bar{f}_{fib}}{1 - \bar{f}_{col}} \tag{D.51}$$

With this value at hand, the mineral volume fractions in the fibrillar [Fig. D.2(b)] and the extrafibrillar space [Fig. D.2(c)] are,

$$\breve{f}_{HA} = \frac{\bar{f}_{HA}(1 - \phi_{HA,ef})}{\bar{f}_{fib}} \tag{D.52}$$

$$\check{f}_{HA} = \frac{\phi_{HA,ef}\bar{f}_{HA}}{\bar{f}_{ef}} \tag{D.53}$$

see Table D.3 for values used to validate the herein proposed strength model.

Within the fibril, comprising the phases hydroxyapatite and wet collagen, the volume fraction of the latter reads as

$$\breve{f}_{wetcol} = 1 - \breve{f}_{HA} \tag{D.54}$$

Finally, the volume fraction of (molecular) collagen at the wet collagen level [Fig. D.2(a)] can be calculated from $\bar{f}_{col}$, through

$$\mathring{f}_{col} = \frac{\bar{f}_{col}}{\breve{f}_{wetcol}} \tag{D.55}$$

see Table D.3 for values used for validating the proposed strength model.

D.6.2 Experimental set providing tissue-specific strength values for model testing

In most cases, strength of bone is quantified in terms of uniaxial, compressive or tensile mechanical tests, under quasi-static conditions (i.e. with a strain rate well below one). To show the relevance of our model approach, we consider various experimental results from various laboratories and various test setups, on various different bone samples (see Table D.4 for specimen geometries, employed machines, and strain rates, and Table D.5 for tissue-specific experimental results).

D.6.3 Comparison between tissue-specific strength predictions and corresponding experiments

The strength values predicted by the six-step homogenization scheme (Fig. D.2) for tissue-specific volume fractions (experimental set of Subsection D.6.1) on the basis of tissue-independent

literature source	specimen geometry [mm]	machine	strain rate [1/s]
(Burstein et al. 1972)	cylindrical (d_S=5) with rcs (d_{cs}=2.9)	not given	not given
(Burstein et al. 1975)	cuboidal (≈15x5x5) with rcs (a=2)	not given	not given
(Burstein et al. 1976)	cuboidal (≈15x5x5) with rcs (a=2)	not given	0.05
(Cezayirlioglu et al. 1985)	cuboidal (4-5x4x45) with rcs (d_{cs}=2.5-3)	Instron 1230	0.01-0.06
(Currey 1959)	cylindrical (l_S=28) with rcs (d_{cs}=1.9-2.7)	not given	not given
(Currey 1975)	cuboidal with rcs (a_{cs}=1.8)	Instron table model	1.3x10^{-4}-0.16
(Currey 1990)	cuboidal with rcs (a_{cs}=1.8)	Instron 1122	0.2
(Currey 2004)	cuboidal with rcs (a_{cs}=1.8)	Instron 1122	0.2
(Dickenson et al. 1981)	cylindrical (l=30, d_S=5.5) with rcs (d_{cs}=2.4)	hydraulic servo-controlled	not given
(Hellmich et al. 2006)	cylindrical (l_S=10, d_S=5)	LFM 150, Wille Geotechnik	0.001
(Kotha and Guzelsu 2002)	cuboidal with rcs (2x5)	Instron	0.0005
(Lee et al. 1997)	cylindrical (l_S=40, d_S=4.5) with rcs (d_{cs}=3)	Instron 1331	0.5
(Martin and Ishida 1989)	cuboidal (45x18x5) with rcs (a_{cs}=5)	Instron 1122	0.004
(McCalden et al. 1993)	cuboidal (32x5x5) with rcs (a_{cs}=2)	J.J. Lloyd M30K	0.03
(Reilly and Burstein 1974a)	cuboidal (≈15x5x5) with rcs (a=2)	not given	0.05
(Reilly and Burstein 1975)	cuboidal (≈15x5x5) with rcs (a=2)	not given	0.02-0.05
(Riggs et al. 1993)	cuboidal (l_S <10) with rcs (tension), cubes (l_S=8, compression)	Instron 6025	0.001
(Sedlin and Hirsch 1966)	cuboidal (≈50x5x2) with rcs	Instron TT-CM	not given

Table D.4: Specimen geometries, employed testing machines, and strain rates of the tensile and compressive tests, see also Table D.5. d_S is the diameter of the sample with length l_S, 'rcs' stands for reduced cross section with diameter d_{cs} or side length a_{cs}.

'universal' phase stiffness and strength properties (experimental set of Tables D.1 and D.2) are compared to corresponding experimentally determined tissue-specific uniaxial tensile and compressive strength values from the experimental set of Subsection D.6.2. The experimental strength values of Subsection D.6.2 are grouped into types of tissues (e.g. human tibia), and their corresponding weighted mean and standard deviation is considered (see Tables D.6 and D.7 as well as Fig. D.5).

To quantify the model's predictive capabilities, we consider the mean and the standard deviation of the relative error between strength predictions and experiments,

$$\bar{e} = \frac{1}{n}\sum e_i = \frac{1}{n}\sum \frac{\Sigma^{ult}_{cort} - \Sigma^{ult}_{exp}}{\Sigma^{ult}_{cort}} \qquad \text{(D.56)}$$

$$e_S = \left[\frac{1}{n-1}\sum (e_i - \bar{e})^2\right]^{\frac{1}{2}} \qquad \text{(D.57)}$$

The satisfactory agreement between model predictions and experiments is quantified by prediction errors of $+2.61 \pm 24.7\%$ for uniaxial tensile strength, and of $-4.00 \pm 8.42\%$ for uniaxial

literature source	tissue	tension		compression	
		n	$\Sigma_{exp}^{ult,t}$ [MPa]	n	$\Sigma_{exp}^{ult,c}$ [MPa]
(Burstein et al. 1972)	bovine femur	25	172	?	283
(Burstein et al. 1975)	bovine tibia	10	188		
(Burstein et al. 1976)	human femur	178	132	95	192
(Burstein et al. 1976)	human tibia	123	155	38	192
(Cezayirlioglu et al. 1985)	human femur	37	136	19	206
(Cezayirlioglu et al. 1985)	human tibia	13	158	9	213
(Cezayirlioglu et al. 1985)	bovine femur	27	162	25	217
(Currey 1959)	bovine femur	46	106.0		
(Currey 1975)	bovine femur	35	124.5		
(Currey 1990)	bovine femur	4	148		
(Currey 1990)	bovine tibia	4	146		
(Currey 2004)	human femur	4	165.7		
(Currey 2004)	bovine femur	10	142.4		
(Dickenson et al. 1981)	human femur	29	117		
(Hellmich et al. 2006)	bovine tibia			3	180
(Kotha and Guzelsu 2002)	bovine femur	9	106.6		
(Lee et al. 1997)	human tibia	11	77.0		
(Martin and Ishida 1989)	bovine femur	10	112		
(McCalden et al. 1993)	human femur	38	91.6		
(Reilly and Burstein 1974a)	human femur	101	128.5	95	192.5
(Reilly and Burstein 1974a)	bovine femur	11	133.1	10	249.6
(Reilly and Burstein 1974a)	bovine tibia	152	228		
(Reilly and Burstein 1975)	human femur	21	135	20	205
(Reilly and Burstein 1975)	bovine femur	3	144	3	272
(Riggs et al. 1993)	equine radius	40	161	13	185
(Riggs et al. 1993)	equine radius	40	105	13	217
(Sedlin and Hirsch 1966)	human femur	52	87.5		

Table D.5: Tissue-specific experimental uniaxial tensile and compressive mean strength values. n denotes the number of samples tested.

tissue	model $\Sigma_{cort}^{ult,t}$ [MPa]	experiments $\Sigma_{exp}^{ult,t}$ mean±std.dev. [MPa]
human femur	122.59	122.59 ± 17.28
human tibia	124.82	149.43 ± 20.69
bovine femur	147.69	132.77 ± 24.75
bovine tibia	125.00	164.00 ± 18.33
equine radius	118.91	133.00 ± 28.18

Table D.6: Predicted and experimental strength values for different tissues tested in uniaxial tension.

compressive strength [$\bar{e} \pm e_S$ according to Eqs. (D.56) and (D.57)].

D.7 Discussion of model characteristics

Having successfully shown the predictive capabilities of the proposed model for various cortical bone tissues tested in uniaxial tension and compression, it is interesting to study the sequence

tissue	model $\Sigma^{ult,c}_{cort}$ [MPa]	experiments $\Sigma^{ult,c}_{exp}$ mean±std.dev. [MPa]
human femur	-187.60	-194.50 ± 5.00
human tibia	-190.84	-196.02 ± 8.35
bovine femur	-246.57	-231.28 ± 20.59
bovine tibia	-197.83	-214.91 ± 22.42
equine radius	-190.19	-201.00 ± 10.81

Table D.7: Predicted and experimental strength values for different tissues tested in uniaxial compression.

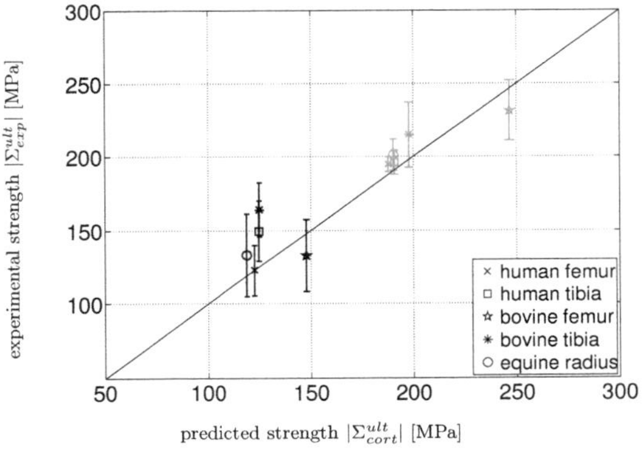

Figure D.5: Comparison between model predictions and experiments at the macroscopic scale [cortical bone material, Fig. D.2(f)]. Mean and standard deviation are depicted for experimental tensile strength (dark color) and experimental compressive strength (light color).

of plastic (interfacial) events in the extrafibrillar space, in terms of the orientations of involved hydroxyapatite crystals.

Under uniaxial tensile loading of cortical bone in axial (longitudinal) direction ($\vartheta = 0°$), longitudinally oriented crystals are the first to undergo inelastic deformation. In the course of further loading, inelastic deformations spread relatively quickly over the range defined by orientation angles ϑ between zero and 30 degrees [see Figure D.6(c)-(e) for $E_{cort,33}$ below 0.1%]. Afterwards, the spreading of plasticity slows off, and stops at an orientation angle of about 65 degrees, see Figure D.6(d)-(e) for plastic strains, and Figure D.6(c) for orientation $\vartheta = 74.25°$ remaining in the elastic regime. Thereby, crystals with longitudinal orientation carry tensile normal stresses at a constant level throughout the plastic loading stage, whereas the normal stresses in inclined crystals are declining, while increasing shear stresses build up [see Figure D.6(a)-(b)].

Under uniaxial compressive loading of cortical bone material in axial (longitudinal) direction ($\vartheta = 0°$), transversely oriented crystals (i.e. such oriented perpendicular to the longitudinal direction) are the first to undergo inelastic deformation. In the course of further loading, inelastic deformations spread relatively quickly over the range defined by orientation angles between 90 and 70 degrees [see Figure D.7(c)-(e) for $E_{cort,33}$ below 0.1%]. Afterwards, the spreading of plasticity slows off, and stops at an orientation angle of about 60 degrees, see Figure D.7(d)-(e) for plastic strains, and Figure D.7(c) for orientation $\vartheta < 60°$ remaining in the elastic regime. Thereby, transversely oriented crystals and crystals with slight inclination from the transverse directions (which are first associated with plasticity) carry normal tensile stresses, while more inclined crystals are loaded in normal compression. Throughout the plastic loading stage, all these crystals, whether loaded normally in tension or in compression, carry increasing shear stresses [see Figure D.7(a)-(b)].

This sequence of plastic events leads to distinctive stress-strain relationships at the level of cortical bone (see Figure D.8): Elastoplastic behavior associated to longitudinal extrafibrillar crystals under tensile loading provokes a decrease of slope in the stress-strain curve, which is more pronounced than that related to elastoplastic behavior in transverse crystal clusters under compression. Thereby, Figure D.8 illustrates the stress-strain curves until the failure stress in the collagen according to (D.24) is reached - this agrees well with the investigations of Pidaparti et al. (1997); Morgan et al. (2005), showing a rather (quasi-)brittle behavior of cortical bone under uniaxial loading. On the other hand, several investigators (Currey 1959; Reilly and Burstein 1974a; Kotha and Guzelsu 2002) report increasing cortical strains at a constant cortical stress level close to the ultimate strength level, i.e. the occurrence of (macroscopically apparent) 'plastic' events also beyond the point when the collagen failure criterion (D.24) is reached in the framework of our model. The micromechanical consideration of respective plastic or microcracking/crack bridging events (as dealt with by various researchers (Burr et al. 1998; Reilly and Currey 2000; Akkus and Rimnac 2001; Okumura and Gennes 2001; Taylor 2003; Ballarini et al. 2005; O'Brien et al. 2007; Koester et al. 2008)) is beyond the scope of this manuscript, where we focus on a model which can predict, as function of the bone sample's composition, the ultimate stress which is bearable by that sample.

It is also interesting to study the effect of species, individual, and organ-specific bone microstructures, on the cortical strength of corresponding bone materials: In healthy mammalian cortical bone, the vascular porosity varies typically between 2 and 8%, while osteoporosis may lead to porosities up to 27% (Bousson et al. 2000). Influence of vascular porosity increase on cortical strength is illustrated in Figures D.9 and D.10, it is of linear nature.

Within the extravascular matrix of a specific organ of an adult mammal, the average chemical composition is constant in space and time (Hellmich et al. 2008), as can be seen from experimental results from computerized contact microradiography (Boivin and Meunier 2002), quantitative backscattered electron imaging (Roschger et al. 2003), Raman microscopy (Akkus et al. 2003), and Synchrotron Micro Computer Tomography (Bossy et al. 2004). Therefore,

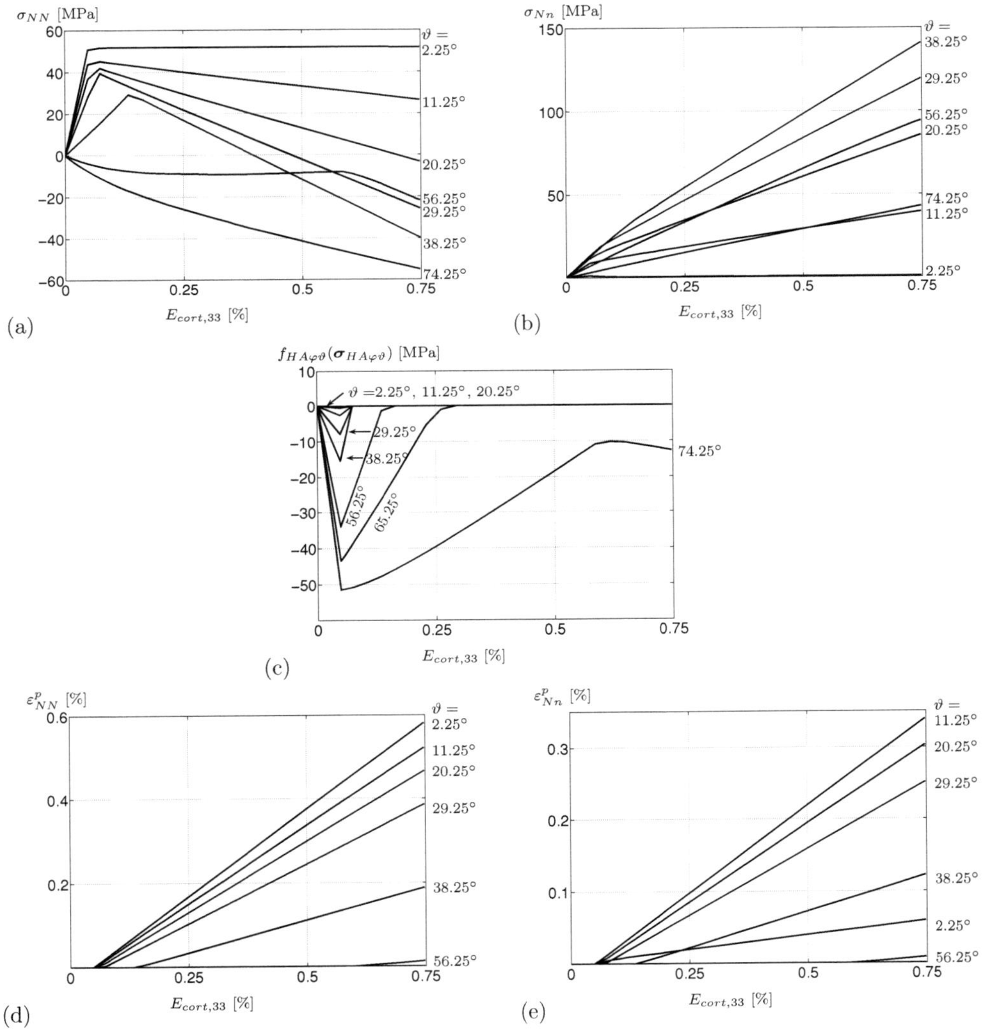

Figure D.6: Plastic mechanisms associated to differently oriented crystals in extrafibrillar space, provoked by uniaxial tensile loading of cortical bone material (human femur, see Table D.3, line 1): (a) normal stress and (b) shear stress; (c) value of yield function; (d) normal plastic strain and (e) shear plastic strain

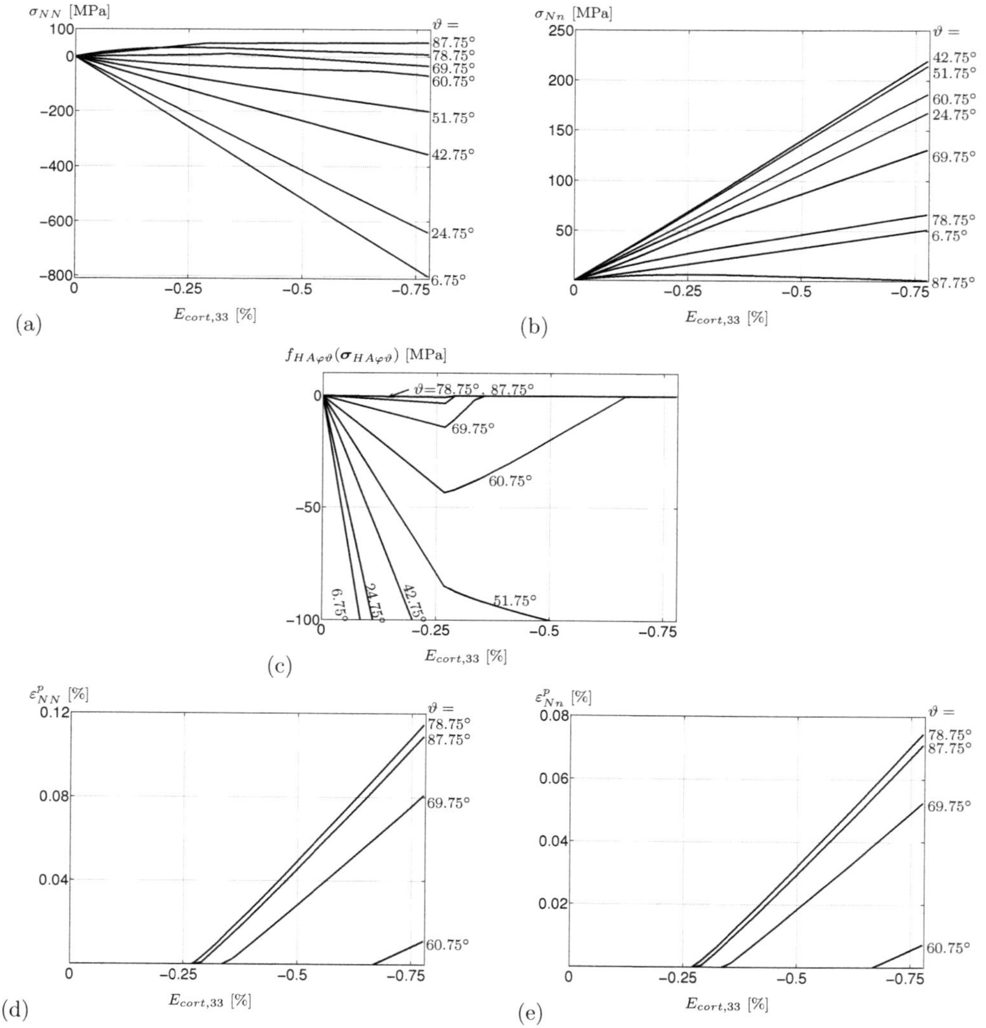

Figure D.7: Plastic mechanisms associated to differently oriented crystals in extrafibrillar space, provoked by uniaxial compressive loading of cortical bone material (human femur, see Table D.3, line 1): (a) normal stress and (b) shear stress; (c) value of yield function; (d) normal plastic strain and (e) shear plastic strain

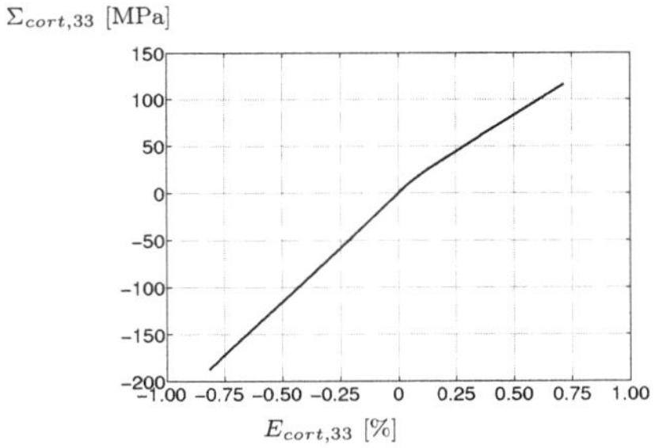

Figure D.8: Macrosopic stress-strain diagram for human femur in uniaxial tension and compression.

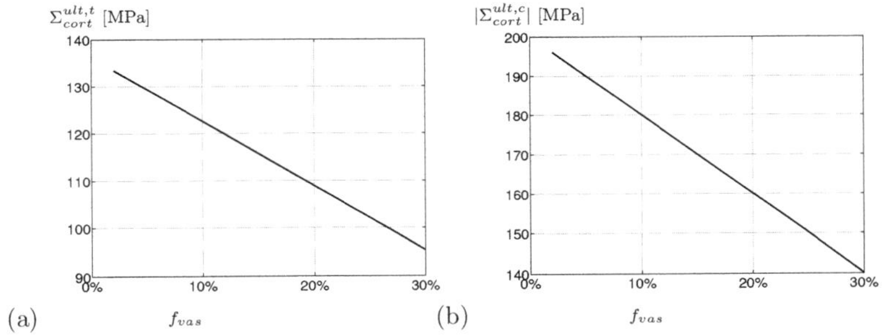

Figure D.9: Model predicted macroscopic uniaxial tensile (a) and compressive (b) strength as function of vascular porosity f_{vas}, for $\bar{f}_{HA} = 46\%$ (human femur).

effects of (varying) extravascular mineral content [while the collagen content follows (D.48)] on different resulting cortical strength values (see Figure D.10), reflect inter-organ and inter-species variations from one bone sample to another, with mineral contents between 30% (typical for deer antler) and 70% (typical for equine metacarpus): the mineralization varying by a factor of two, implies a strength variation by a factor of two in tension, and by a factor of three in compression (Figure D.10). In contrast to the extravascular porosity, the mineral content has a nonlinear influence on cortical strength - this qualitative model feature is in perfect agreement with a wealth of experimental data (Currey 1984, 1988; Hernandez et al. 2001).

Finally, there could seem to be a contradiction between the ductile behavior of interfaces between the hydroxyapatite crystals as part of natural collagenous bone tissue considered in this paper, and the brittle behavior of the interfaces between crystals of man-made hydroxyapatite biomaterials (Akao et al. 1981; Fritsch et al. 2009a). The reason for the different behaviors may well lie in the characteristic size of the crystals, and hence of the nature of their contact surfaces, the crystals in collagenous bone tissue being much smaller than the biomaterial crystals. In the same sense, in low or non-collagenous tissues, such as specific whale bones (Zioupos et al.

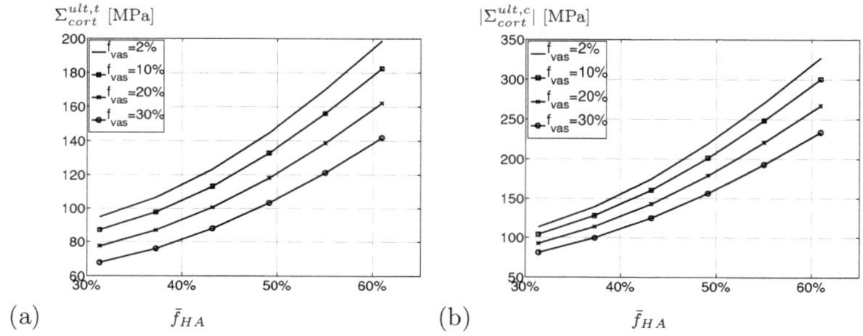

Figure D.10: Model predicted macroscopic uniaxial tensile (a) and compressive (b) strength as function of ultrastructural mineral volume fraction $\bar{f}_{HA}$, for different vascular porosities f_{vas}.

1997), the minerals grow larger, and also these tissues exhibit a brittle failure behavior. The idea of increased ductility due to increased activity of layered water films is also supported by the fact (Nyman et al. 2008) that bound water content is correlated to bone toughness; and this idea fits well with the suggestions of Boskey (2003), that larger crystals (implying less layered water films per crystal content) would lead to a more brittle behavior of bone materials.

D.8 Conclusion and Perspectives

We have proposed a first multiscale micromechanics model for bone strength, extending earlier developments in the realm of elasticity (Hellmich et al. 2004a; Fritsch and Hellmich 2007). Thereby, the explanation of bone strength across different species and ages required resolution of the mineral phase into an infinite amount of non-spherical phases, and definition of an elastoplastic failure criterion for the mineral crystals, reflecting layered water-induced ductile sliding between these mineral crystals. The multiscale material model was validated through independent experimental results: Tissue-specific strength values predicted by the micromechanical model on the basis of tissue-independent ('universal') stiffnesses and strengths of the elementary components (mineral, collagen, water), for tissue-specific composition data (volume fractions) were compared to corresponding experimentally determined tissue-specific strength values. Mean relative errors between stiffness experiments and model estimates were well below 10%, which, given remarkable experimental scattering, is considered satisfactory.

This renders the model ready for supporting various future scientific as well as application-oriented activities:

1. As was already shown for elasticity (Hellmich et al. 2008), the model is expected to be combined with computer tomographic images: Based on average relations from X-ray

physics, the voxel-specific X-ray attenuation information would be translated to voxel-specific material composition; and the latter would serve as input for the micromechanical model, which would then deliver voxel-specific (anisotropic and inhomogeneous) stiffness *and* strength values. In this way, the current activities concerning the virtual physiological human (Taylor et al. 2002; Yosibash et al. 2007; Viceconti et al. 2008), could be extended from the realm of elasticity to that of full elastoplasticity, resulting in patient-specific fracture risk assessment of whole organs in both healthy and pathological conditions.

2. The proposed model could also support the design of tissue engineering scaffolds, through predictions of the failure properties of bone tissue-engineering scaffolds with tissue-engineered bone, by feeding recently developed multiscale representations (Bertrand and Hellmich 2008) not only with an elastic, but with the present elastoplastic micromechanical representation of the extracellular bone material.

3. Since the proposed model is linked to the hierarchical organization of bone and to its elementary components, it is ready to be combined with most recent developments in theoretical and computational biochemistry and biology, which quantify the well-tuned interplay of biological cells via biochemical signaling pathways (Lemaire et al. 2004; Pivonka et al. 2008) – giving as output the volume fraction of newly deposited or resorbed extravascular bone, which may serve as input for the proposed multiscale strength model. That is expected to open the way to translation of biochemical remodeling events to associated changes in mechanical competence.

D.9 Appendix: Hill tensors $\mathbb{P}$

D.9.1 Hill tensor for homogenization over wet collagen

$\mathbb{P}^{col}_{cyl}$ refers to a cylindrical inclusion in a transversely isotropic matrix with stiffness $\mathbb{c}_{col}$, where the plane of isotropy is oriented perpendicular to the long axis of the cylinder. The non-zero components of the symmetric tensor $\mathbb{P}^{col}_{cyl}$ read as follows (Hellmich et al. 2004a; Levin et al. 2000):

$$P^{col}_{cyl,1111} = P^{col}_{cyl,2222} = 1/8\,(5\,c_{col,1111} - 3\,c_{col,1122})/c_{col,1111}/\mathcal{D}_2\,, \tag{D.58}$$

$$P^{col}_{cyl,1122} = P^{col}_{cyl,2211} = -1/8\,(c_{col,1111} + c_{col,1122})/c_{col,1111}/\mathcal{D}_2\,, \tag{D.59}$$

$$P^{col}_{cyl,2323} = P^{col}_{cyl,1313} = 1/(8\,c_{col,2323})\,, \tag{D.60}$$

$$P^{col}_{cyl,1212} = 1/8\,(3\,c_{col,1111} - c_{col,1122})/c_{col,1111}/\mathcal{D}_2\,, \tag{D.61}$$

whereby

$$\mathcal{D}_2 = c_{col,1111} - c_{col,1122} \tag{D.62}$$

D.9.2 Hill tensors for homogenization over mineralized collagen fibril

The non-zero components of $\mathbb{P}^{fib}_{cyl}$ follow from substitution of '$c_{col,ijkl}$' by '$C^{SCS}_{fib,ijkl}$' in (D.58)-(D.62). The non-zero components of $\mathbb{P}^{fib}_{sph}$ for spherical inclusions in a transversely isotropic matrix follow from substitution of 'C^0_{ijkl}' by '$C^{SCS}_{fib,ijkl}$' in the following equations:

$$\begin{aligned} P^0_{sph,1111} = \frac{1}{16}\int_{-1}^{1} &-(-5C^0_{1111}x^4C^0_{3333} - 3C^0_{1122}x^2C^0_{3333} - 3C^0_{1122}x^4C^0_{2323} \\ &+3C^0_{1122}x^4C^0_{3333} + 5C^0_{1111}x^4C^0_{2323} - 10C^0_{1111}C^0_{2323}x^2 + 2x^4C^{0,2}_{1133} \\ &+8C^0_{2323}x^4C^0_{3333} - 6C^{0,2}_{2323}x^4 + 4C^0_{2323}x^4C^0_{1133} + 6C^0_{1122}C^0_{2323}x^2 \\ &+5C^0_{1111}C^0_{2323} + 5C^0_{1111}x^2C^0_{3333} - 4C^0_{2323}x^2C^0_{1133} + 6C^{0,2}_{2323}x^2 \\ &-2x^2C^{0,2}_{1133} - 3C^0_{1122}C^0_{2323})(-1+x^2)/\mathcal{D}_1 dx \end{aligned} \tag{D.63}$$

$$\begin{aligned} P^0_{sph,1122} = P^0_{sph,2211} = \frac{1}{16}\int_{-1}^{1} &(C^0_{1111}C^0_{2323} - 2C^0_{1111}C^0_{2323}x^2 + C^0_{1111}x^2C^0_{3333} \\ &+C^0_{1122}C^0_{2323} - 2C^0_{1122}C^0_{2323}x^2 + C^0_{1122}x^2C^0_{3333} + C^0_{1111}x^4C^0_{2323} - C^0_{1111}x^4C^0_{3333} \\ &+C^0_{1122}x^4C^0_{2323} - C^0_{1122}x^4C^0_{3333} - 2C^{0,2}_{2323}x^2 + 2C^{0,2}_{2323}x^4 - 4C^0_{2323}x^2C^0_{1133} \\ &+4C^0_{2323}x^4C^0_{1133} - 2x^2C^{0,2}_{1133} + 2x^4C^{0,2}_{1133})(-1+x^2)/\mathcal{D}_1 dx \end{aligned} \tag{D.64}$$

$$P^0_{sph,1133} = P^0_{sph,3311} = \frac{1}{4}\int_{-1}^{1}(-1+x^2)x^2(C^0_{2323} + C^0_{1133})/\mathcal{D}_2 dx \tag{D.65}$$

$$P^0_{sph,2323} = \frac{1}{16}\int_{-1}^{1}(4C^0_{1111}C^0_{2323}x^2 - 8C^0_{2323}x^4C^0_{1133} - 2x^4C^{0,2}_{1133} - C^0_{1122}x^4C^0_{3333}$$
$$-8C^0_{1111}x^4C^0_{2323} + 3C^0_{1111}x^4C^0_{3333} + 4C^0_{1111}x^4C^0_{1133} - 4C^0_{1122}x^4C^0_{1133}$$
$$+2C^0_{1122}x^6C^0_{1133} - 2C^0_{1111}x^6C^0_{1133} + C^0_{1122}x^6C^0_{1111} - 3C^0_{1122}x^4C^0_{1111}$$
$$+3C^0_{1122}C^0_{1111}x^2 - 2C^0_{1111}x^2C^0_{1133} + 2C^0_{1122}x^2C^0_{1133} + 8x^6C^0_{2323}C^0_{1133}$$
$$-3x^6C^0_{1111}C^0_{3333} + 4x^6C^0_{2323}C^0_{3333} + 4C^0_{1111}x^6C^0_{2323} + C^0_{1122}x^6C^0_{3333} + 3C^{0,2}_{1111}x^4$$
$$-C^{0,2}_{1111}x^6 + 2C^{0,2}_{1133}x^6 - 3C^{0,2}_{1111}x^2 + C^{0,2}_{1111} - C^0_{1122}C^0_{1111})/\mathcal{D}_1 dx \tag{D.66}$$

$$P^0_{sph,3333} = \frac{1}{2}\int_{-1}^{1} x^2(x^2C^0_{2323} - C^0_{1111}x^2 + C^0_{1111})/\mathcal{D}_2 dx \tag{D.67}$$

whereby

$$\mathcal{D}_1 = -2C^{0,2}_{1111}x^4C^0_{3333} + 2C^{0,2}_{2323}x^6C^0_{3333} - 4C^0_{1111}C^{0,2}_{2323}x^4 - 3C^{0,2}_{1111}C^0_{2323}x^2 + C^{0,2}_{1111}x^2C^0_{3333} +$$
$$2C^0_{1111}C^{0,2}_{2323}x^2 - 2C^0_{2323}x^4C^{0,2}_{1133} - C^0_{1111}C^{0,2}_{1133}x^6 + 2C^0_{1111}C^{0,2}_{1133}x^4 + 4C^{0,2}_{2323}x^6C^0_{1133}$$
$$-2C^0_{1122}C^{0,2}_{1133}x^4 + 2C^0_{2323}x^6C^{0,2}_{1133} + 3C^{0,2}_{1111}x^4C^0_{2323} + C^0_{1122}C^{0,2}_{1133}x^6 - C^{0,2}_{1111}x^6C^0_{2323}$$
$$+2C^0_{1111}x^6C^{0,2}_{2323} + C^{0,2}_{1111}x^6C^0_{3333} - C^0_{1111}C^{0,2}_{1133}x^2 - 4C^{0,2}_{2323}x^4C^0_{1133} + C^0_{1122}C^{0,2}_{1133}x^2$$
$$+C^{0,2}_{1111}C^0_{2323} - C^0_{1122}C^0_{1111}C^0_{2323} - C^0_{1122}x^6C^0_{1111}C^0_{3333} + 4C^0_{1111}x^4C^0_{2323}C^0_{1133} - 2C^0_{1111}x^2C^0_{2323}C^0_{1133}$$
$$-4C^0_{1122}x^4C^0_{2323}C^0_{1133} + 2C^0_{1122}x^2C^0_{2323}C^0_{1133} + 2C^0_{1122}x^6C^0_{2323}C^0_{1133} - 2C^0_{1111}x^6C^0_{2323}C^0_{1133}$$
$$-3C^0_{1111}x^6C^0_{2323}C^0_{3333} + 2C^0_{1122}C^0_{1111}x^4C^0_{3333} - C^0_{1122}C^0_{2323}x^4C^0_{3333} - 3C^0_{1122}C^0_{1111}x^4C^0_{2323}$$
$$-C^0_{1122}C^0_{1111}x^2C^0_{3333} + 3C^0_{1122}C^0_{1111}C^0_{2323}x^2 + 3C^0_{1111}C^0_{2323}x^4C^0_{3333} + C^0_{1122}x^6C^0_{1111}C^0_{2323}$$
$$+C^0_{1122}x^6C^0_{2323}C^0_{3333} \tag{D.68}$$

and

$$\mathcal{D}_2 = 2C^0_{2323}x^4C^0_{1133} + C^0_{2323}x^4C^0_{3333} + C^0_{1111}x^4C^0_{2323} - 2C^0_{2323}x^2C^0_{1133} - 2C^0_{1111}C^0_{2323}x^2$$
$$+C^0_{1111}C^0_{2323} + x^4C^{0,2}_{1133} - C^0_{1111}x^4C^0_{3333} - x^2C^{0,2}_{1133} + C^0_{1111}x^2C^0_{3333} \tag{D.69}$$

D.9.3 Hill tensors for homogenization over extrafibrillar space

$\mathbb{P}^{ef}_{sph}$, the Hill tensor for a spherical inclusion in an isotropic matrix of stiffness $\mathbb{C}^{SCSII}_{ef}$, is of the form (Eshelby 1957; Zaoui 1997b)

$$\mathbb{P}^{ef}_{sph} = \mathbb{S}^{esh,ef}_{sph} : \mathbb{C}^{SCSII,-1}_{ef}, \tag{D.70}$$

$$\mathbb{S}^{esh,ef}_{sph} = \alpha^{SCSII}_{ef}\,\mathbb{J} + \beta^{SCSII}_{ef}\,\mathbb{K} \tag{D.71}$$

with

$$\alpha_{ef}^{SCSII} = \frac{3\,k_{ef}^{SCSII}}{3\,k_{ef}^{SCSII} + 4\,\mu_{ef}^{SCSII}}$$
$$\beta_{ef}^{SCSII} = \frac{6\,(k_{ef}^{SCSII} + 2\,\mu_{ef}^{SCSII})}{5\,(3\,k_{ef}^{SCSII} + 4\,\mu_{ef}^{SCSII})} \tag{D.72}$$

$\mathbb{P}_{cyl}^{ef}$, the Hill tensor for a cylindrical inclusion in an isotropic matrix, is of the form

$$\mathbb{P}_{cyl}^{ef} = \mathbb{S}_{cyl}^{esh} : \mathbb{C}_{ef}^{SCSII,-1} \tag{D.73}$$

The non-zero components of the Eshelby tensor $\mathbb{S}_{cyl}^{esh}$ corresponding to cylindrical inclusions read as

$$S_{cyl,1111}^{esh} = S_{cyl,2222}^{esh} = \frac{5 - 4\nu_{ef}^{SCSII}}{8(1 - \nu_{ef}^{SCSII})}$$
$$S_{cyl,1122}^{esh} = S_{cyl,2211}^{esh} = \frac{-1 + 4\nu_{ef}^{SCSII}}{8(1 - \nu_{ef}^{SCSII})}$$
$$S_{cyl,1133}^{esh} = S_{cyl,2233}^{esh} = \frac{\nu_{ef}^{SCSII}}{2(1 - \nu_{ef}^{SCSII})}$$
$$S_{cyl,2323}^{esh} = S_{cyl,3232}^{esh} = S_{cyl,3223}^{esh} = S_{cyl,2332}^{esh} =$$
$$= S_{cyl,3131}^{esh} = S_{cyl,1313}^{esh} = S_{cyl,1331}^{esh} = S_{cyl,3113}^{esh} = \frac{1}{4}$$
$$S_{cyl,1212}^{esh} = S_{cyl,2121}^{esh} = S_{cyl,2112}^{esh} = S_{cyl,1221}^{esh} = \frac{3 - 4\nu_{ef}^{SCSII}}{8(1 - \nu_{ef}^{SCSII})} \tag{D.74}$$

where principal directions 1, 2, and 3 follow Figure D.2, and with ν_{ef}^{SCSII} as Poisson's ratio of the extrafibrillar space,

$$\nu_{ef} = \frac{3k_{ef}^{SCSII} - 2\mu_{ef}^{SCSII}}{6k_{ef}^{SCSII} + 2\mu_{ef}^{SCSII}} \tag{D.75}$$

Following standard tensor calculus (Salencon 2001), the tensor components of $\mathbb{P}_{cyl}^{ef}(\vartheta, \varphi)$, being related to differently oriented inclusions, are transformed into one, single base frame ($\underline{e}_1$, $\underline{e}_2$, $\underline{e}_3$), in order to evaluate the integrals in Eq. (D.27).

D.9.4 Hill tensor for homogenization over extracellular bone matrix

$\mathbb{P}_{cyl}^{ef}$, the Hill tensor for a cylindrical inclusion in an isotropic matrix, is given in Eq. D.73.

D.9.5 Hill tensor for homogenization over extravascular bone material

The non-zero components of $\mathbb{P}^{excel}_{sph}$ for spherical inclusions in a transversely isotropic matrix follow from substitution of 'c^0_{ijkl}' by '$C^{MTII}_{excel,ijkl}$' in Eqs (D.63)-(D.69).

D.9.6 Hill tensor for homogenization over cortical bone material

The non-zero components of $\mathbb{P}^{exvas}_{cyl}$ for cylindrical inclusions in a transversely isotropic matrix follow from substitution of '$c_{col,ijkl}$' by '$C^{MTIII}_{exvas,ijkl}$' in Eqs (D.58)-(D.62).

Nomenclature

a_{cs}	side length of reduced cross section of a bone specimen
$\mathbb{C}_{rs}$	fourth-order influence tensor
$\mathcal{A}$	constant in the linear relationship between ρ_{excel} and $\bar{f}_{HA}$
$\mathbb{A}_r$	fourth-order strain concentration tensor of phase r
b	width of a volume of one rhomboidal fibrillar unit
$\mathcal{B}$	constant in the linear relationship between ρ_{excel} and $\bar{f}_{HA}$
$\mathbb{C}_{col}$	fourth-order stiffness tensor of molecular collagen
$c_{col,ijkl}$	component of fourth-order stiffness tensor of molecular collagen
$\mathcal{C}$	constant in the linear relationship between ρ_{excel} and d_s
$\mathbb{C}_{cort}^{MTIV}$	homogenized fourth-order stiffness tensor of cortical bone material
$\mathbb{C}_{ef}^{SCSII}$	homogenized fourth-order stiffness tensor of extrafibrillar space
$\mathbb{C}_{excel}^{MTII}$	homogenized fourth-order stiffness tensor of extracellular bone matrix
$\mathbb{C}_{exvas}^{MTIII}$	homogenized fourth-order stiffness tensor of extravascular bone material
$\mathbb{C}_{fib}^{SCS}$	homogenized fourth-order stiffness tensor of mineralized collagen fibril
$\mathbb{C}_{HA}$	fourth-order stiffness tensor of hydroxyapatite
$\mathbb{C}_{ic}$	fourth-order stiffness tensor of intercrystalline space
$\mathbb{C}_{im}$	fourth-order stiffness tensor of intermolecular water
$\mathbb{C}_{inc}$	fourth-order stiffness tensor of an inclusion embedded in a matrix with stiffness $\mathbb{C}^0$
$\mathbb{C}_{lac}$	fourth-order stiffness tensor of lacunae
$\mathbb{C}_M$	fourth-order stiffness tensor of the matrix phase
$\mathbb{C}_r$	fourth-order stiffness tensor of phase r
$\mathbb{C}_{vas}$	fourth-order stiffness tensor of Haversian canals
$\mathbb{C}_{wetcol}^{MT}$	homogenized fourth-order stiffness tensor of wet collagen
$\mathbb{C}^{hom}$	homogenized fourth-order stiffness tensor
$\mathbb{C}^0$	fourth-order stiffness tensor of an infinite matrix surrounding an ellipsoidal inclusion
d	characteristic length of the inhomogeneities within an RVE
d_{cs}	diameter of reduced cross section of a bone specimen
d_s	neutron diffraction spacing between collagen molecules
d_S	diameter of a bone specimen
D	1/5 of length of a volume of one rhomboidal fibrillar unit
$\mathcal{D}$	constant in the linear relationship between ρ_{excel} and d_s
$\boldsymbol{E}$	second-order 'macroscopic' strain tensor
$\boldsymbol{E}_r$	second-order 'macroscopic' strain tensor of phase r
$\boldsymbol{E}_{r,n}$, $\boldsymbol{E}_{r,n+1}$	second-order 'macroscopic' strain tensors of phase r for load steps n and $n+1$, respectively
$\boldsymbol{E}_{r,n}^p$, $\boldsymbol{E}_{r,n+1}^p$	second-order 'macroscopic' plastic strain tensors of phase r for load steps n and $n+1$, respectively

$\boldsymbol{E}^{p(k)}_{r,n+1}$	k-th approximation of second-order 'macroscopic' plastic strain tensor of phase r for load step $n+1$
$\boldsymbol{E}^{trial}_{r,n+1}$	second-order 'macroscopic' trial strain tensor of phase r for load step $n+1$
$\boldsymbol{E}^{p}$	second-order 'macroscopic' plastic strain tensor
$\boldsymbol{E}^{0,p}$	uniform 'macroscopic' plastic strain in matrix of a matrix-inclusion problem
$\boldsymbol{E}^{\infty}$	uniform 'macroscopic' strain at infinity of a matrix-inclusion problem
$\underline{e}_1, \underline{e}_2, \underline{e}_3$	unit base vectors of Cartesian reference base frame
$\underline{e}_\vartheta, \underline{e}_\varphi, \underline{e}_r$	unit base vectors of Cartesian local base frame of a single crystal of hydroxyapatite within extrafibrillar space
$\mathfrak{f}_r(\boldsymbol{\sigma}_r)$	boundary r of elastic domain of phase r in space of microstresses
$\bar{f}_{col}$	volume fraction of collagen within an RVE $\bar{V}_{excel}$
$\mathring{f}_{col}$	volume fraction of molecular collagen within an RVE $\mathring{V}_{wetcol}$
$\bar{f}_{ef}$	volume fraction of extrafibrillar space within an RVE $\bar{V}_{excel}$
$\tilde{f}_{excel}$	volume fraction of extracellular bone matrix within an RVE $\tilde{V}_{exvas}$
f_{exvas}	volume fraction of extravascular bone material within an RVE V_{cort}
$\bar{f}_{fib}$	volume fraction of mineralized collagen fibril within an RVE $\bar{V}_{excel}$
$\bar{f}_{HA}$	volume fraction of hydroxyapatite within an RVE $\bar{V}_{excel}$
$\breve{f}_{HA}$	volume fraction of hydroxyapatite within an RVE $\breve{V}_{fib}$
$\check{f}_{HA}$	volume fraction of hydroxyapatite within an RVE $\check{V}_{ef}$
$\bar{f}_{H_2O}$	volume fraction of water within an RVE $\bar{V}_{excel}$
$\check{f}_{ic}$	volume fraction of intercrystalline space within an RVE $\check{V}_{ef}$
$\mathring{f}_{im}$	volume fraction of intermolecular water within an RVE $\mathring{V}_{wetcol}$
$\tilde{f}_{lac}$	volume fraction of lacunae within an RVE $\tilde{V}_{exvas}$
$\bar{f}_{org}$	volume fraction of organic matter within an RVE $\bar{V}_{excel}$
f_r	volume fraction of phase r
f_{vas}	volume fraction of Haversian canals within an RVE V_{cort}
$\breve{f}_{wetcol}$	volume fraction of wet collagen within an RVE $\breve{V}_{fib}$
HA	hydroxyapatite
$\mathbb{I}$	fourth-order identity tensor
$\mathbb{J}$	volumetric part of fourth-order identity tensor $\mathbb{I}$
$\mathbb{K}$	deviatoric part of fourth-order identity tensor $\mathbb{I}$
k_{HA}	bulk modulus of hydroxyapatite
k_{H_2O}	bulk modulus of water
$\mathcal{L}$	characteristic lengths of geometry or loading of a structure built up by the material defined on the RVE
l_S	length of a bone specimen
ℓ	characteristic length of an RVE
ℓ_{cort}	characteristic length of an RVE V_{cort} of cortical bone material

ℓ_{ef}	characteristic length of an RVE $\check{V}_{ef}$ of extrafibrillar space
ℓ_{excel}	characteristic length of an RVE $\bar{V}_{excel}$ of extracellular bone matrix
ℓ_{exvas}	characteristic length of an RVE $\tilde{V}_{exvas}$ of extravascular bone material
ℓ_{fib}	characteristic length of an RVE $\check{V}_{fib}$ of mineralized collagen fibril
ℓ_{wetcol}	characteristic length of an RVE $\mathring{V}_{col}$ of wet collagen
M	index denoting a material phase being the matrix
$\underline{N}$	orientation vector aligned with longitudinal axis of hydroxyapatite needle
n_r	number of material phases within an RVE
$\underline{n}$	orientation vector perpendicular to $\underline{N}$
RVE	representative volume element
r	index denoting a material phase
$\mathbb{P}^0_{inc}$	fourth-order Hill tensor characterizing the interaction between the inclusion *inc* and the matrix $\mathbb{C}^0$
$\mathbb{P}^0_r$	fourth-order Hill tensor characterizing the interaction between the phase r and the matrix $\mathbb{C}^0$
sgn(.)	signum function of quantity (.)
$\mathbb{S}$	fourth-order Eshelby tensor for spherical inclusions
v_{col}	volume of a single collagen molecule
v_{fib}	volume of one rhomboidal fibrillar unit
$\mathring{V}_{col}$	volume of molecular collagen within an RVE $\mathring{V}_{wetcol}$
V_{cort}	volume of RVE 'cortical bone material'
$\check{V}_{ef}$	volume of RVE 'extrafibrillar space'
$\bar{V}_{ef}$	volume of extrafibrillar space within an RVE $\bar{V}_{excel}$
$\bar{V}_{excel}$	volume of RVE 'extracellular bone matrix'
$\tilde{V}_{excel}$	volume of extracellular bone matrix within an RVE $\tilde{V}_{exvas}$
$\tilde{V}_{exvas}$	volume of RVE 'extravascular bone material'
V_{exvas}	volume of extravascular bone material within an RVE V_{cort}
$\check{V}_{fib}$	volume of RVE 'mineralized collagen fibril'
$\bar{V}_{fib}$	volume of mineralized collagen fibril within an RVE $\bar{V}_{excel}$
$\check{V}_{HA}$	volume of hydroxyapatite within an RVE $\check{V}_{fib}$
$\check{V}_{HA}$	volume of hydroxyapatite within an RVE $\check{V}_{ef}$
$\check{V}_{ic}$	volume of intercrystalline space within an RVE $\check{V}_{ef}$
$\mathring{V}_{im}$	volume of intermolecular water within an RVE $\mathring{V}_{wetcol}$
$\tilde{V}_{lac}$	volume of lacunae within an RVE $\tilde{V}_{exvas}$
V_{vas}	volume of Haversian canals within an RVE V_{cort}
$\mathring{V}_{wetcol}$	volume of RVE 'wet collagen'
$\check{V}_{wetcol}$	volume of wet collagen within an RVE $\check{V}_{fib}$
WF^{cort}_{HA}	weight fraction of hydroxyapatite at the scale of cortical bone material
WF^{excel}_{HA}	weight fraction of hydroxyapatite at the extracellular scale
WF^{cort}_{org}	weight fraction of organic matter at the scale of cortical bone material

WF_{org}^{excel}	weight fraction of organic matter at the extracellular scale
β	ratio between uniaxial tensile strength and shear strength of pure HA
$\Delta\boldsymbol{E}_{r,n+1}$	incremental second-order 'macroscopic' strain tensor of phase r for load step $n+1$
$\Delta\boldsymbol{E}_{r,n+1}^{p}$	incremental second-order 'macroscopic' plastic strain tensor of phase r for load step $n+1$
$\Delta\boldsymbol{E}_{r,n+1}^{p(k)}$	k-th approximation of incremental second-order 'macroscopic' plastic strain tensor of phase r for load step $n+1$
$\Delta\boldsymbol{\varepsilon}_{n+1}^{p}$	incrmental plastic strain of $n+1$-st load increment
$\Delta\lambda_{HA,n+1}$	incrmental plastic multiplier of $n+1$-st load increment
$\boldsymbol{\varepsilon}_{col}$	second-order strain tensor field within molecular collagen
$\boldsymbol{\varepsilon}_{ef}$	second-order strain tensor field within an RVE $\check{V}_{ef}$ of extrafibrillar space
$\boldsymbol{\varepsilon}_{excel}$	second-order strain tensor field within an RVE $\bar{V}_{excel}$ of extracellular bone matrix
$\boldsymbol{\varepsilon}_{exvas}$	second-order strain tensor field within an RVE $\tilde{V}_{exvas}$ of extravascular bone material
$\boldsymbol{\varepsilon}_{fib}$	second-order strain tensor field within an RVE $\breve{V}_{fib}$ of mineralized collagen fibril
$\boldsymbol{\varepsilon}_{HA\vartheta\varphi}$	second-order strain tensor field within oriented hydroxyapatite needles in extrafibrillar space
$\boldsymbol{\varepsilon}_{inc}$	second-order strain tensor field within an inclusion embedded in matrix $\mathbb{C}^0$
$\boldsymbol{\varepsilon}_{inc}^{p}$	second-order plastic strain tensor field within an inclusion embedded in matrix $\mathbb{C}^0$
$\bar{\varepsilon}_{ij}$	tensor component of difference $(\boldsymbol{\varepsilon}_{HA\varphi\vartheta,n+1} - \boldsymbol{\varepsilon}_{HA\varphi\vartheta,n}^{p})$, given in a local base frame
$\boldsymbol{\varepsilon}_{M}^{p}$	second-order plastic strain tensor field within the matrix phase
$\boldsymbol{\varepsilon}_{n}^{p}$, $\boldsymbol{\varepsilon}_{n+1}^{p}$	second-order strain tensor fields for load steps n and $n+1$, respectively
$\boldsymbol{\varepsilon}_{r}$	second-order 'microscopic' strain tensor field within phase r
$\dot{\boldsymbol{\varepsilon}}_{r}$	incremental 'microscopic' second-order strain tensor field within phase r
$\boldsymbol{\varepsilon}_{r}^{p}$	second-order 'microscopic' plastic strain tensor field within phase r
$\boldsymbol{\varepsilon}_{r}^{trial}$	second-order 'microscopic' trial strain tensor field within phase r
$\boldsymbol{\varepsilon}_{wetcol}$	second-order strain tensor field within an RVE $\mathring{V}_{col}$ of wet collagen
$\dot{\lambda}_{r}$	incremental plastic multiplier
ϑ	latitudinal coordinate of spherical coordinate system
θ	integration variable, $\theta = 0 \ldots \pi$
μ_{HA}	shear modulus of hydroxyapatite
μ_{H_2O}	shear modulus of water
ρ_{col}	mass density of molecular collagen
ρ_{cort}	mass density of cortical bone material
ρ_{excel}	mass density of the extracellular bone matrix

ρ_{HA}	mass density of hydroxyapatite
ρ_{H_2O}	mass density of water
ρ_{org}	mass density of organic matter
$\boldsymbol{\sigma}_{col}$	second-order stress tensor field within molecular collagen
σ_{col}^{ult}	uniaxial tensile or compressive strength of molecular collagen
$\boldsymbol{\sigma}_{ef}$	second-order stress tensor field within an RVE $\check{V}_{ef}$ of extrafibrillar space
$\boldsymbol{\sigma}_{excel}$	second-order stress tensor field within an RVE $\bar{V}_{excel}$ of extracellular bone matrix
$\boldsymbol{\sigma}_{exvas}$	second-order stress tensor field within an RVE $\tilde{V}_{exvas}$ of extravascular bone material
$\boldsymbol{\sigma}_{fib}$	second-order stress tensor field within an RVE $\breve{V}_{fib}$ of mineralized collagen fibril
$\boldsymbol{\sigma}_{HA\vartheta,\varphi}$	second-order stress tensor field within oriented hydroxyapatite needle in extrafibrillar space
σ_{HA}^{NN}	normal component of stress tensor $\boldsymbol{\sigma}_{HA\vartheta\varphi}$ in needle direction
σ_{HA}^{Nn}	shear component of stress tensor $\boldsymbol{\sigma}_{HA\vartheta\varphi}$ in planes orthogonal to the needle direction
$\boldsymbol{\sigma}_{HA\vartheta\varphi,n+1}^{trial}$	second-order trial stress tensor field within oriented HA needle for load step $n+1$
$\sigma_{HA}^{ult,s}$	uniaxial shear strength of pure HA
$\sigma_{HA}^{ult,t}$	uniaxial tensile strength of pure HA
$\boldsymbol{\sigma}_r$	second-order stress tensor field within phase r
$\boldsymbol{\sigma}_r^{(k)}$	k-th approximation of stress field within phase r
$\boldsymbol{\sigma}_{wetcol}$	second-order stress tensor field within an RVE $\mathring{V}_{col}$ of wet collagen
$\boldsymbol{\Sigma}$	second-order 'macroscopic' stress tensor
$\boldsymbol{\Sigma}_{cort}$	second-order stress tensor within an RVE V_{cort} of cortical bone material
Σ_{cort}^{ult}	model-predicted uniaxial strength of cortical bone material
Σ_{exp}^{ult}	experimental uniaxial strength of cortical bone material
φ	longitudinal coordinate of spherical coordinate system
ϕ	integration variable, $\phi = 0..2\pi$
$\phi_{HA,ef}$	relative amount of hydroxyapatite in the extrafibrillar space
ψ	longitudinal coordinate of vector $\underline{n}$
$\cdot$	first-order tensor contraction
$:$	second-order tensor contraction
$\otimes$	dyadic product of tensors

Publication E

Acoustical and poromechanical characterization of titanium scaffolds for biomedical applications (Müllner et al. 2008)

Authored by Herbert W. Müllner, Andreas Fritsch, Christoph Kohlhauser, Roland Reihsner, Christian Hellmich, Dirk Godlinski, Astrid Rota, Robert Slesinski, and Josef Eberhardsteiner
Published in *Strain*, Volume 44, pages 153–163

Biocompatible materials are designed so as to mimic biological materials such as bone as closely as possible. As regards the mechanical aspect of bone replacement materials, a certain stiffness and strength are mandatory to effectively carry the loads imposed on the skeleton. In this paper, porous titanium with different porosities, produced on the basis of metal powder and space holder components, is investigated as bone replacement material. For the determination of mechanical properties, i.e. strength of dense and porous titanium samples, two kinds of experiments were performed - uniaxial and triaxial tests. The triaxial tests were of poromechanical nature, i.e. oil was employed to induce the same pressure both at the lateral surfaces of the cylindrical samples and inside the pores. The stiffness properties were revealed by acoustic (ultrasonic) tests. Different frequencies give access to different stiffness components (stiffness tensor components related to high-frequency-induced bulk waves versus Young's moduli related to low-frequency-induced bar waves), at different observation scales; namely, the observation scale the dense titanium with around 100 μm characteristic length (characterized through the high frequencies) versus that of the porous material with a few millimetres of characteristic length (characterized through the low frequencies). Finally, the experimental results were used to develop and validate a poro-micromechanical model for porous titanium, which quantifies material stiffness and strength from its porosity and (in the case of the aforementioned triaxial tests) its pore pressurisation state.

Notation

a	radius of cylindrical specimen
$\mathbb{C}_{hom}$	Homogenized stiffness of porous medium
$\mathbb{C}_S$	Elasticity tensor of pure titanium
C_{1111}	Normal component of isotropic elasticity tensor
C_{1212}	Shear component of isotropic elasticity tensor
d	Characteristic size of inhomogeneities within material volume (RVE)
div	divergence of a vector field
$\boldsymbol{e}_{1,2,3}$	Base vectors
$\boldsymbol{E}$	Macroscopic strain tensor
E	Young's modulus of porous titanium
E_S	Young's modulus of pure titanium
$\mathfrak{F}$	Homogenized, macroscopic yield criterion
f	Frequency
f_y	Yield stress
G	Shear modulus of porous titanium
i	Index denoting tensor components
$\mathbb{I}$	Fourth-order identity tensor
j	Index denoting tensor components
$\mathbb{J}$	Volumetric part of fourth-order identity tensor
J_0	Bessel function of first kind and order 0
J_1	Bessel function of first kind and order 1
$\mathbb{K}$	Deviatoric part of fourth-order identity tensor
k_f	Compressibility of porous medium
k_S	Bulk modulus of pure titanium
ℓ_{RVE}	Characteristic length of the RVE
l_S	Travel distance through the specimen
p	Pore pressure in porous titanium
p_0	Lateral pressure built up in pressure cell
m	Fluid mass per unit volume of porous medium
r	Radial polar coordinate
RVE	Representative volume element
$\mathbb{S}$	Eshelby tensor
t	Time
tr	trace of tensor
t_S	Travel time through the specimen
v	Phase velocity of acoustic wave

v_L	Bulk velocity of longitudinal (or compressional) wave
v_{bar}	Bar velocity of bar wave
v_T	Velocity of transversal (or shear) wave
$\boldsymbol{v}$	Fluid velocity
V_{solid}	Solid volume inside the RVE of porous medium
$\boldsymbol{w}$	Mass flow vector
$\boldsymbol{x}$	Location vector in the RVE
$\mathbf{1}$	Second-order identity tensor
α_n	Roots of J_0, $J_0(\alpha_n) = 0$
β	Inverse characteristic time of surface pressue built-up
δ_{ij}	Kronecker delta
Δ	Laplace operator
$\boldsymbol{\varepsilon}$	Microscopic strain tensor
ε_d	Equivalent (micro-) shear strains
$\varepsilon_{eff,d}$	Effective equivalent deviatoric microstrains
η_f	Viscosity of fluid
κ	Intrinsic permeability of porous medium
λ	Wavelength
μ_S	Shear modulus of pure titanium
ν	Poisson's ratio of porous titanium
ν_S	Poisson's ratio of pure titanium
ρ	Mass density of specimen
ρ_f	Mass density of fluid
$\boldsymbol{\Sigma}$	Macroscopic stress tensor
Σ_d	Equivalent deviatoric macroscopic stress
Σ_m	Mean macroscopic stress
φ	Porosity of porous medium
$:$	Second-order tensor contraction
$\otimes$	Dyadic product of tensors

E.1 Introduction

Many bone replacement materials, based on a multitude of different chemical compositions, are available nowadays. All these materials are designed so as to mimic bone as closely as possible. In other words, the bone biomaterials are required to be biocompatible (Jones 2005), i.e. they should smoothly fit into the biological, chemical, and mechanical environment inside the body of the patient. As regards the mechanical aspect, a certain stiffness and strength are mandatory to effectively carry the loads imposed onto the skeleton. In addition, the biomaterial should match the mechanical properties of the original bone as precisely as possible, in order to preserve the standard physiological stress fields around the implant. These stress fields are required to guarantee effective functioning of the biological cells resorbing the bone and forming new bone.

In this study,we aimed at contributing to the latter aspect. Precise determination of the stress fields around an implant requires profound knowledge of the material properties of both the bone material and the bone replacement material under multiaxial stress states, as found in the living body (Kobayashi et al. 2001). In addition to multiaxial stress fields, the pore pressure inside the bone is often believed to play a mandatory role, as regards both mechanical integrity (Hellmich and Ulm 2005a,b; Ochoa et al. 1991; Lim and Hong 2000) and biological function (Mizuno et al. 2004; Weinbaum et al. 1994). However, related experimental data are extremely scarce in the open literature. Therefore, we have started a campaign of triaxial test series on bone and bone biomaterials, giving access to the strength properties of the tested materials. Moreover, to determine the stiffness of such materials, our test campaign included ultrasonic measurements as well. Here we describe processing as well as its mechanical and acoustic characterization of titanium biomaterials. Finally, the experimental results are used to develop and validate a first poro-micromechanical model for porous titanium, which quantifies material stiffness and strength from its porosity and (in the case of the aforementioned triaxial tests) its pore pressurization state.

E.2 Materials

Porous titanium samples with open cell structures were produced by using metal powder (pure titanium particles with <45μm characteristic length) and spherical space holder components (para-formaldehyde with a mean diameter of 500 μm), at Fraunhofer IFAM (Bremen, Germany). The manufacturing process included four steps.

1. *Powder mixture preparation:* Titanium and para-formaldehyde (as space holder) were mixed with paraffin (as a pressing agent), and with additional process aids dissolved in water or organic solvent, to ensure a good bonding of the metal powder and the space holder particles.

2. *Pressing:* The mixture was densified, by means of axial pressing in a powder press.

3. *Debinding:* After compaction, the space holder and bonding agent phases was removed from the samples, in a catalytic process.

4. *Sintering:* After complete space holder removal, the samples were sintered in a high vacuum atmosphere, at a temperature of 1200°C.

The above-described process ensures crack-free and homogeneous titanium samples, with two different porosities (Figure E.1).

1. *Dense titanium* [Figure E.1(a), background] was processed without space holders. However, the formation of some microns-sized pores inside the material (Figure E.2) results in

Figure E.1: (a) Titanium samples (porous in foreground, dense in background); (b) higher magnification of porous titanium samples

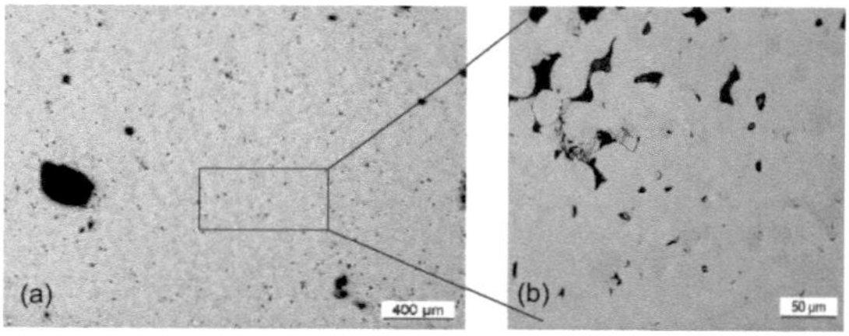

Figure E.2: (a) Micrograph of the center of a dense titanium sample; (b) higher magnification of the denser part of the same sample

a mass density of 3.80 g/cm^3, remarkably lower than the mass density of pure titanium, which is 4.50 g/cm^3 (Thelen et al. 2004).

2. *Porous titanium* [Figure E.1(a), foreground; and Figure E.1(b)] was produced by use of spaceholders as described before. The solid matrix between the hundreds-of-microns-sized pores is similar to the material depicted in Figure E.2. The overall porous material exhibits a mass density of 1.64 g/cm^3.

E.3 Mechanical testing

All tests were conducted at room temperature. The average height and diameter of the samples were 10.0 and 5.0 mm, respectively. In uniaxial testing mode, the samples were subjected to axial compressive loads by means of a 150 kN uniaxial electromechanical machine [LFM 150; Wille Geotechnik, Germany, with displacement control, Figure E.3(a)], at a displacement rate of 0.01 mm/s. Extension of uniaxial testing mode to triaxial loading was realized through a high-pressure triaxial testing cell [LT 63500-2/50-T; Wille Geotechnik, Germany, Figure E.3(b)], filled with mineral oil. In order to stabilize the sample during the filling process, it was attached to the lower die by means of plasticine [Figure E.3(d)].

An outlet valve on the top of the cell eliminated air bubbles within the testing chamber. This valve was locked once the chamber was properly filled with oil. Then, the oil was pressurized by means of an electromechanical pressure control [DV 350-150/10; Wille Geotechnik, Figure E.3(c)], up to a pressure of 14.5 MPa. Pressures of this order of magnitude occur if the bone is deformed under undrained conditions (Lim and Hong 2000). A vertical compressive force was applied simultaneously by the electromechanical uniaxial testing machine. The specimens were loaded in a state of axisymmetric triaxial compressive stress until the vertical displacement of the upper die [Figure E.3(d)], driven by the electromechanical machine, reached 30% of the specimen height.

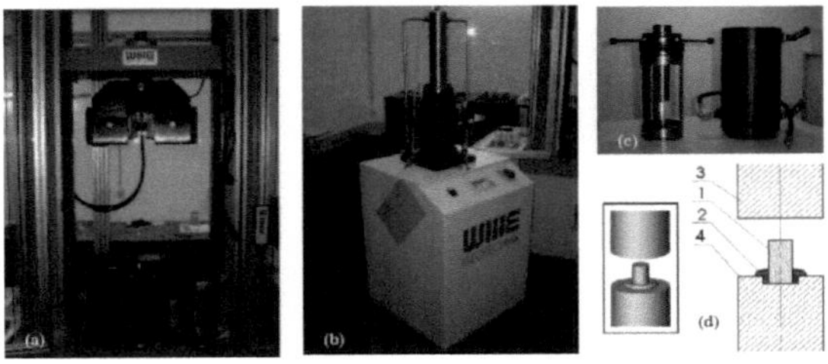

Figure E.3: Experimental setup for uniaxial and triaxial tests: (a) 150 kN uniaxial testing machine; (b) pressure control; (c) 150 bar triaxial cell; (d) fixing of specimen: (1) specimen, (2) plasticine, (3) upper die, (4) lower die

E.3.1 Identification of triaxial tests as poromechanical tests

Here, we show that the pore pressure build-up within the porous titanium samples is very much faster than the uniaxial load application through the electromechanical machine, so that the uniaxial macroscopic deformation is increased, while a constant pore pressure is prescribed in the pores. In order to estimate corresponding characteristic times, we study the transport of oil through an undeformed (incompressible) porous medium (metal foam).

The fluid mass conservation law for this case reads as

$$\frac{\mathrm{d}m}{\mathrm{d}t} + \mathrm{div}\, \boldsymbol{w} = 0, \tag{E.1}$$

where m is the fluid mass per unit volume of porous medium, $\mathrm{d}(.)/\mathrm{d}t$ denotes the temporal derivation of quantity (.), div denotes the divergence of a vector field, and $\boldsymbol{w}$ is the mass fluid vector. The latter is related to the fluid velocity $\boldsymbol{v}$ through

$$\boldsymbol{w} = \varphi \rho_f \boldsymbol{v}, \tag{E.2}$$

where φ is the porosity and ρ_f the mass density of the fluid. The fluid mass change is related to the fluid pressure change $\mathrm{d}p/\mathrm{d}t$ through the state equation of the fluid (Coussy 2004)

$$\frac{\mathrm{d}m}{\mathrm{d}t} = \varphi \frac{\mathrm{d}\rho_f}{\mathrm{d}t} = \varphi\,\rho_f \frac{1}{k_f}\frac{\mathrm{d}p}{\mathrm{d}t}, \tag{E.3}$$

where $k_f = 1.5$ GPa (Rydberg 2001) is the compressibility or bulk modulus of the (oil) fluid. The fluid velocity $\boldsymbol{v}$ results from a pressure gradient, as expressed in Darcy's fluid conduction law

$$\boldsymbol{v} = -\frac{\kappa}{\eta_f}\,\mathrm{grad}\,p, \tag{E.4}$$

where η_f is the fluid viscosity ($\eta_f = 450$ mPas for oil (Grimm and Williams 1997)), and κ the intrinsic permeability of the porous medium ($\kappa = 3.1$ x 10^{-8} m^2 for an open metal foam of comparable porosity (Leong and Jin 2006)). Use of Equations (E.2)-(E.4) in (E.1) yields an analogon to the so-called diffusion equation (Crank 1975), reading for space-invariant material properties k_f, η_f and κ, as

$$\frac{\mathrm{d}p}{\mathrm{d}t} = \frac{k_f\kappa}{\eta_f}\Delta p, \tag{E.5}$$

with Δ as the Laplace operator.

Solutions of this partial differential equation are widely documented, see e.g. (Crank 1975). Specifically, the pore pressure development $p(r,t)$ inside a cylindrical porous sample due to rapid pressure build-up around the sample,

$$p = p_0(1-\exp(-\beta t)) \quad \text{with } \beta \rightarrow \infty \tag{E.6}$$

can be given in the form (Crank 1975):

$$\frac{p}{p_0} = 1 - \frac{J_0(\sqrt{\beta r^2 \eta_f/k_f\kappa})}{J_0(\sqrt{\beta a^2 \eta_f/k_f\kappa})}\exp(-\beta t) + \frac{2\beta\eta_f}{a k_f\kappa}\sum_{n=1}^{\infty}\frac{J_0(r\alpha_n)}{\alpha_n J_1(a\alpha_n)}\frac{\exp(-k_f\kappa\alpha_n^2 t/\eta_f)}{\alpha_n^2-(\beta\eta_f/k_f\kappa)} \tag{E.7}$$

where r is the radial polar coordinate, t denotes the time elapsed since the initiation of pressure build-up, J_0 and J_1 are the Bessel functions of the first kind and of order 0 and 1, respectively, and α_n are the roots of J_0, $J_0(\alpha_n) = 0$: $\alpha_1 = 2.4048$, $\alpha_2 = 5.5201$, $\alpha_3 = 8.6537$, ...

Evaluation of Eq. (E.7) for the intrinsic permeability values of metal foams (Table E.1), and the compressibility and viscosity of mineral oil, $k_f = 1.5$ GPa (Rydberg 2001) and $\eta_f = 450$ mPas (McNeil and Stuart 2004), respectively, clearly shows that the pore pressure inside the tested titanium samples is built up within a small fraction of 1 s. This holds even for the intrinsic permeability values of bone (Table E.1) which are lower than the one for metal foams. Hence, during the mechanical experiments, lasting typically 10 min, the pore pressure is always quasi-identical to the oil pressure built up in the pressure cell. Therefore, the triaxial tests performed here may be regarded as poromechanical tests, where the pore pressure *inside* the samples is prescribed.

Source	Material	κ (m^2)
(Leong and Jin 2006)	Metal foam	3.1 x 10^{-8}
(Grimm and Williams 1997)	Trabecular bone	8.5 x 10^{-9}
(Li et al. 1987)	Cortical bone	2.5 x 10^{-13}

Table E.1: Intrinsic permeabilities κ of metal foams and bone

E.3.2 Determination of strength properties

Load-displacement curves obtained for uniaxial and triaxial tests (Figure E.4) are characterized by a considerable decrease of the slope of the load-displacement curve at a certain load level. This refers to ductile material behavior, which is also evident from the deformed shape of the samples after mechanical testing, as shown in the photographs of Figure E.5. Bilinear approximation of the load-displacement curves gives access to the yield load (Figure E.4). Dividing the latter by the sectional area of the specimen gives access to the yield stress of the material (see Table E.2 for corresponding experimental results). The results of the uniaxial and triaxial tests are not markedly different. This is probably due to the fact that the lateral pressure of 14.5 MPa is by far smaller than the uniaxial yield stress of the samples. More profound investigations into the poromechanical behavior of the titanium materials considered herein would call for a pressure cell apt for extremely high pressures.

The remarkably high ductility of the titanium materials does not necessarily match the mechanical characteristics of natural bone, often showing a more brittle behavior in compression (Morgan et al. 2005). This underlines the fact that, in addition to the anisotropy of natural bone (Lees et al. 1979b), which is not mimicked by the tested biomaterial, the inelastic constitutive behavior of man-made biomaterials still needs to be improved as to match more precisely the one of natural bone.

The load-displacement curves presented in Figure E.4 do not show any linear regime, which indicates that inelastic phenomena are at action right from the initial testing phase, when they are restricted to the regions of the samples close to the load platens. Hence, elastic properties cannot be derived from the load-displacement curves; therefore, the mechanical tests were used for determination of strength properties, only; and the materials' elasticity was revealed through ultrasonics measurements (shown below).

	Titanium dense	Titanium porous
Uniaxial test ($p = 0$ MPa)	400 $\pm$ 26 (n=4)	103 $\pm$ 32 (n=4)
Triaxial test ($p = 15$ MPa)	353 $\pm$ 70 (n=4)	88 $\pm$ 15 (n=4)

Table E.2: Mean values and standard deviations of yield stresses in [MPa] ($p \ldots$ oil pressure, $n \ldots$ number of tests)

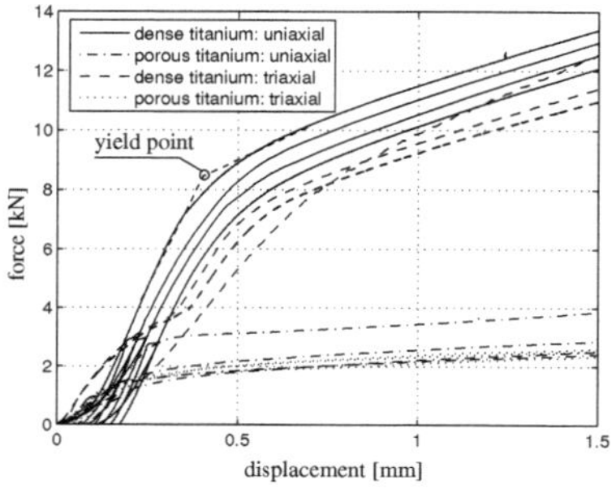

Figure E.4: Load-displacement curves for dense and for porous titanium samples

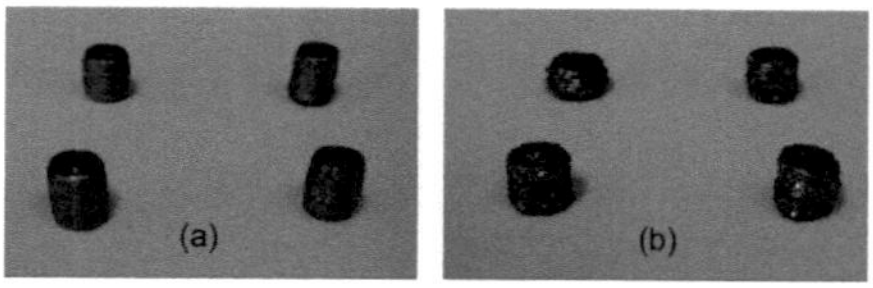

Figure E.5: Photographs of tested samples: (a) dense titanium; (b) porous titanium

E.4 Acoustical Testing

E.4.1 Equipment for transmission through technique

The used ultrasonic device consists of a pulser-receiver PR 5077 [Panametrics Inc., Waltham, MA, Figure E.6(a)], an oscilloscope, and several ultrasonic transducers [Figure E.6(b)]. The pulser unit emits an electrical square-pulse of up to 400 V, with frequencies from 0.1 MHz to 20 MHz. The piezoelectric elements inside the ultrasonic transducers transform such electrical signals into mechanical signals [when operating in the sending mode, transferring, via a coupling medium (here honey), the mechanical signals to one side of the specimen under investigation], or they transform mechanical signals back to electrical signals (when receiving mechanical signals from the opposite side of the specimen under investigation). The piezoelectric elements are tailored for the frequency of the employed mechanical signal: The higher the frequency, the smaller the element and the corresponding transducer. Depending on the cut and orientation of the element, a longitudinal or a transversal wave is emitted.

The receiver unit of the pulser-receiver has a bandwidth of 0.1 to 35 MHz and a voltage gain of up to 59 dB. The amplified signal is displayed on an oscilloscope Lecroy WaveRunner 62Xi (Lecroy Corporoation, Chestnut Ridge, NY) width a bandwidth of 600 MHz and a sample

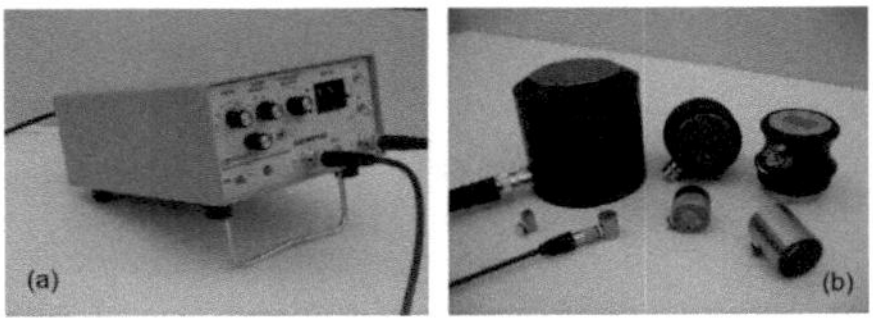

Figure E.6: Equipment for acoustical testing: (a) pulser-receiver; (b) ultrasonic transducers

rate of 10 gigasamples per second. The oscilloscope gives access to the time of flight of the ultrasonic wave through the specimen, t_S, which provides, together with the travel distance through the specimen, l_S, the phase velocity of the wave as

$$v = \frac{l_S}{t_S} \tag{E.8}$$

see Table E.3 for typical velocities of longitudinal or compressional waves (v_L), where the particle displacement points into the wave propagation direction, and transverse or shear waves (v_T), where the particle displacement is perpendicular to the wave propagation direction.

	ρ [g/cm^3]	f [MHz]	v [km/s]	λ [mm]	ℓ_{RVE} [mm]	C_{1111}/C_{1212} [GPa]	E/G [GPa]	ν
Dense	3.83±0.05	10.0	v_L= 5.59±0.02	0.56±0.00	≥0.10	C_{1111}= 119.7±2.3	E= 94.3±4.0	0.28±0.03
Dense	3.83±0.05	5.0	v_T= 3.11±0.12	0.62±0.02	≥0.10	C_{1212}= 37.0±2.3	G= 37.0±2.3	
Dense	3.83±0.05	0.1	v_{bar}= 5.06±0.09	50.6±0.9	≥0.10		E= 98.1±4.4	
Porous	1.69±0.09	0.1	v_{bar}= 3.39±0.05	33.9±0.5	≥2.50		E= 19.5±1.7	

Table E.3: Ultrasonic measurement results for dense and porous titanium samples (mean values ± standard deviations)

E.4.2 Theoretical basis of ultrasonic measurements

Frequency f and wave velocity v give access to the wavelength λ, through

$$\lambda = \frac{v}{f} \tag{E.9}$$

If the wavelength is considerably smaller than the diameter of the specimen, a (compressional) 'bulk wave', i.e. a laterally constrained wave, propagates with velocity v_L in a quasi-infinite medium. On the other hand, if the wavelength is considerably larger than the diameter of the specimen, a 'bar wave' propagates with velocity v_{bar}, i.e. the specimen acts as one-dimensional bar without lateral constraints (Ashman et al. 1984). In contrast, shear waves' propagation is identical in quasi-infinite media and bar-like structures (Ashman et al. 1987).

As regards bulk waves, a combination of the conservation law of linear momentum, the generalized Hooke's law, the linearized strain tensor, and the general plane wave solution for the displacements inside an infinite solid medium yields the elasticity tensor components C_{1111} and C_{1212} of isotropic materials as functions of the material mass density ρ and the bulk wave propagation velocities v_L and v_T (Carcione 2001),

$$C_{1111} = \rho v_L^2 \quad \text{and} \quad C_{1212} = G = \rho v_T^2 \tag{E.10}$$

with G as the shear modulus.

Combination of (E.10) with the definitions of the engineering constants Young's modulus E and Poisson's ratio ν, yields the latter as functions of the wave velocities, in the form

$$E = \rho \frac{v_T^2 (3 v_L^2 - 4 v_T^2)}{v_L^2 - v_T^2} \tag{E.11}$$

and

$$\nu = \frac{E}{2G} - 1 = \frac{v_L^2/2 - v_T^2}{v_L^2 - v_T^2} \tag{E.12}$$

respectively.

In the case of bar wave propagation (Kolsky 1953), the measured bar wave velocity v_{bar} gives direct access to the Young's modulus,

$$E = \rho v_{bar}^2 \tag{E.13}$$

In continuum (micro)mechanics (Zaoui 1997b, 2002), elastic properties are related to a material volume [representative volume element (RVE)], with a characteristic length ℓ_{RVE} being considerably larger than the inhomogeneities d inside the RVE, and the RVE being subjected to homogeneous stress and strain states (Figures E.7 and E.8). Hence, the characteristic length of the RVE, ℓ_{RVE}, needs to be much smaller than the scale of the characteristic loading of the medium, here the wavelength λ (Figure E.7). Mathematically,

$$d \ll \ell_{RVE} \ll \lambda \tag{E.14}$$

Therefore, ultrasonic tests at different frequencies 'detect', inside a sample, materials at different observation scales (Fritsch and Hellmich 2007), such as the macroscopic porous material or the solid phase of the material. In the following, this is detailed for the titanium samples.

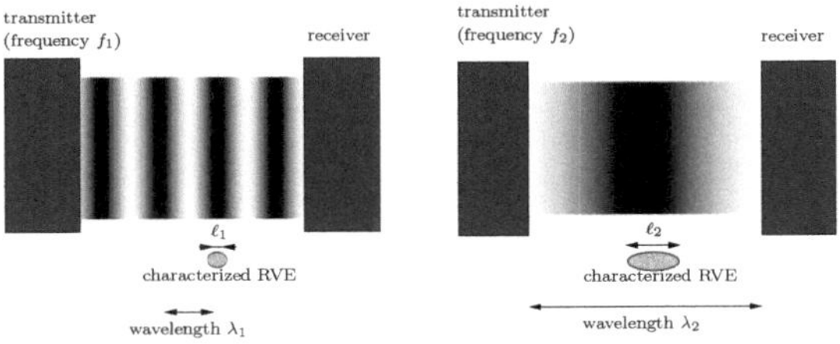

Figure E.7: Schematic, grey-scale based illustration of stress magnitude in specimens tested ultrasonically with different frequencies ($f_1 > f_2$) (Fritsch and Hellmich 2007)

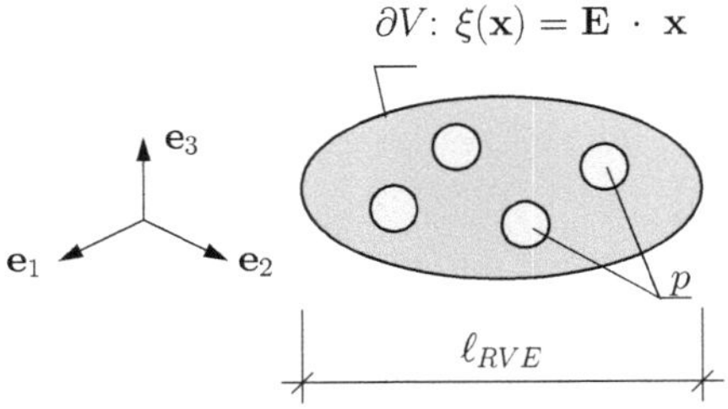

Figure E.8: Micromechanical representation of porous medium (Dormieux 2005; Dormieux et al. 2002, 2006b): a representative volume element (RVE) is loaded by displacements related to homogeneous (macroscopic) strains $\boldsymbol{E}$, and by a pore pressure p

E.4.3 Determination of elastic properties

Longitudinal waves at ultrasonic frequencies of 0.1 and 10 MHz, and transversal waves at 5 MHz were employed to characterize four dense and four porous cylindrical samples. The waves traveled along the height of the specimen.

The employed frequencies implied wavelengths of around half a millimeter and half a decimeter, respectively (Table E.3), characterizing the RVEs of dense and porous titanium samples, with at least 0.1 and 2.5 mm characteristic length, respectively (Table E.3). Depending on the wavelength, measured velocities correspond to bulk waves (rows 1 and 2 of Table E.3) or to bar waves (rows 3 and 4 of Table E.3). Remarkably, two independent test series at different frequencies, providing Young's modulus of dense titanium either directly ($f = 0.1$ MHz) or via C_{1111} and C_{1212} ($f = 5$ and $f = 10$ MHz), differ by only 3% (rows 3 and 1 in Table E.3).

E.5 Prediction of mechanical properties by means of poro-micromechanics – microstructure-property relationships

In this section, we aim at explaining the above-collected stiffness and strength properties from the internal structure and composition of the tested materials. Therefore, we consider the basic morphological feature of the pores inside the samples, which is its spherical shape, and the volume occupied by these pores normalized by the volume of the entire material volume, i.e. the porosity of the samples. In a first micromechanical approximation of the material's microstructure, we do not distinguish between the typically 10-μm-sized pores discernable in Figure E.2 and the typically 500-μm-sized pores discernible in Figure E.1; but we consider the sum of both porosities as overall porosity. Accordingly, the measured mass density of each specimen and the mass density of pure titanium, equal to 4.50 g/cm3, give access to the aforementioned overall porosity of each sample (see coordinates on abscissa of experimental data points in Figures E.9 and E.10, as well as Table E.4 for mean values and standard deviations).

We consider an RVE of porous titanium (Figure E.8, see also Section E.4 and Figure E.7), with characteristic length $\ell_{RVE} = 2 \ldots 5$ mm. Therein, we distinguish two quasi-homogeneous subdomains (also called material phases): (i) the pores of characteristic size $d = 10 \ldots 500$ microns $\ll \ell_{RVE}$, with a volume fraction equal to the porosity φ and with a prescribed hydrostatic stress state equal to the pore pressure; and (ii) the solid titanium matrix with volume fraction $(1-\varphi)$ and with mechanical properties of pure (non-porous) titanium. The elastic properties of the latter are typically given by a Young's modulus $E_S = 120$ GPa and a Poissons ratio $\nu_S = 0.32$, i.e. by a bulk modulus $k_S = 111$ GPa and a shear modulus $\mu_S = 45.5$ GPa (Matweb 2007), see also the stiffnesses in Figure E.9 at $\varphi = 0$, and the uniaxial strength of pure titanium typically amounts to 450 MPa (Matweb 2007). These quantities are the basis for determination of the 'homogenized' mechanical behavior of the overall material, i.e. the relation between homogeneous ('macroscopic') deformations $\boldsymbol{E}$ acting on the boundary of the RVE (being identical to the average of the ('micro'-) strains inside the RVE) and resulting average ('micro'-) stresses (being identical to the 'macroscopic' stresses $\boldsymbol{\Sigma}$), as well as the macroscopic stress states related to material failure ('homogenized strength'). The homogenized or effective material behavior of the porous titanium samples is estimated from the mechanical behavior of the aforementioned homogeneous phases, representing the inhomogeneities within the RVE, their dosages within the RVE, their characteristic shapes, and their interactions, as described next.

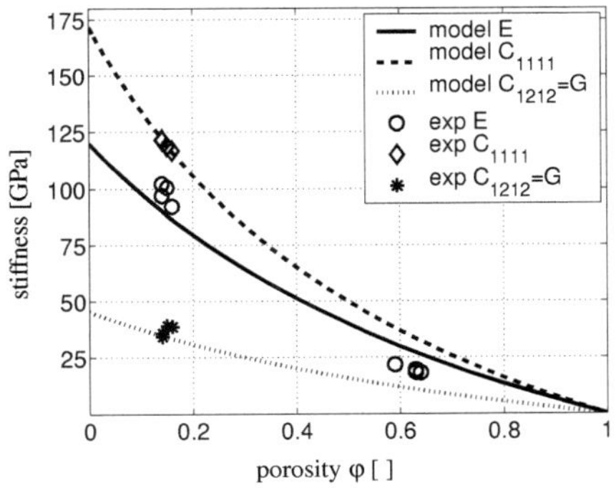

Figure E.9: Prediction of stiffness properties of titanium samples, by means of poro-micromechanical model, Equations (E.15)-(E.18); experimental values according to Sections E.3 and E.4

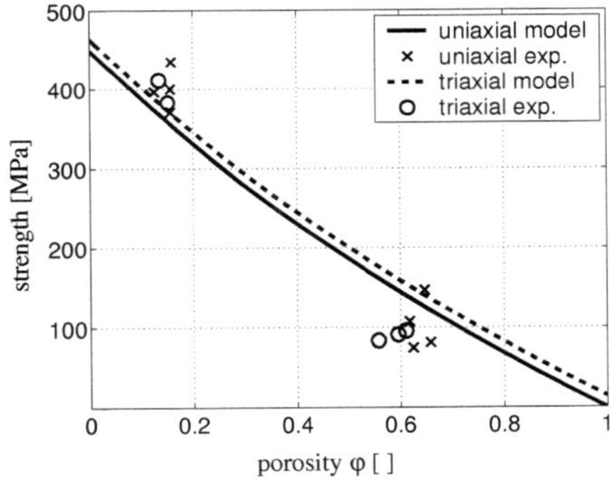

Figure E.10: Prediction of strength properties of titanium samples, by means of poro-micromechanical model, Equations (E.19)-(E.20); experimental values according to Sections E.3 and E.4

E.5.1 Stiffness

For predicting the effective stiffness properties of the (empty) porous titanium samples, we consider - on average - the interaction of spherical pores inside a pure titanium matrix, by

	φ (%)
Dense titanium samples	14.9 ± 1.2
Porous titanium samples	62.4 ± 2.1

Table E.4: Porosities of samples (mean values $\pm$ standard deviations)

means of a Mori-Tanaka homogenization scheme (mean-field homogenization) (Zaoui 2002; Dormieux 2005; Dormieux et al. 2002; Mori and Tanaka 1973; Benveniste 1987), delivering the following estimate $\mathbb{C}_{hom}$ for the 'homogenized' stiffness of the composite material 'porous titanium'

$$\mathbb{C}_{hom} = \mathbb{C}_S : (\mathbb{I} - \varphi[\mathbb{I} - (\mathbf{1} - \varphi)\mathbb{S}]^{-1}), \tag{E.15}$$

relating macroscopic stresses $\mathbf{\Sigma}$ to macroscopic strain $\boldsymbol{E}$. In (E.15), $\mathbb{C}_S$ is the elasticity tensor of pure titanium, $\mathbb{C}_S = 3k_S\mathbb{J} + 2\mu_S\mathbb{K}$ with

$$\mathbb{J} = \frac{1}{3}\mathbf{1} \otimes \mathbf{1} \quad \text{and} \quad \mathbb{K} = \mathbb{I} - \mathbb{J} \tag{E.16}$$

as the volumetric and the deviatoric part of the fourth-order identity tensor,

$$\mathbb{I} = I_{ijkl} = \frac{1}{2}(\delta_{ik}\delta_{jl} + \delta_{il}\delta_{kj}) \tag{E.17}$$

and δ_{ij} (Kronecker delta) are the components of the second-order identity tensor $\mathbf{1}$, δ_{ij}=1 for i=j and 0 otherwise.

The Eshelby tensor $\mathbb{S}$ for spherical inclusions accounts for the inclusion shape and is of the form (Eshelby 1957)

$$\mathbb{S} = \frac{3k_S}{3k_S + 4\mu_S}\mathbb{J} + \frac{6(k_S + 2\mu_S)}{5(3k_S + 4\mu_S)}\mathbb{K} \tag{E.18}$$

The predictions of the micromechanical model (E.15)-(E.18) compare well with corresponding experimentally determined stiffnesses (Figure E.9).

E.5.2 Strength

In contrast to the homogenized elastic properties, which can be derived from averages of microstrains and microstresses over the material phases, homogenization of strength properties calls for additional information on the heterogeneity of these micro-quantities, i.e. the strain or stress peaks inside the microstructure (possibly cancelled out through averaging) need to be appropriately considered.

It has recently been shown (Dormieux et al. 2002; Kreher 1990), that this heterogeneity can reasonably be considered through the so-called effective microstrains, such as the square root of the average over the solid material phase, of the squares of the equivalent deviatoric (micro-)strains $\boldsymbol{\varepsilon}_d(\boldsymbol{x})$,

$$\varepsilon_{eff,d} = \sqrt{\int_{V_{solid}} \boldsymbol{\varepsilon}_d(\boldsymbol{x}) : \boldsymbol{\varepsilon}_d(\boldsymbol{x})\mathrm{d}V} \tag{E.19}$$

with

$$\boldsymbol{\varepsilon}_d(\boldsymbol{x}) = \boldsymbol{\varepsilon}(\boldsymbol{x}) - \frac{1}{3}\mathrm{tr}\boldsymbol{\varepsilon}(\boldsymbol{x})\mathbf{1} \tag{E.20}$$

where V_{solid} is the volume inside the RVE, which is occupied by the solid matrix, $\boldsymbol{x}$ the location vector indicating positions inside the RVE (Figure E.8), and tr denotes the trace of a tensor. By non-linear homogenization theory (Dormieux 2005; Dormieux et al. 2002; Suquet 1997a), the limit case of large effective microstrains, being related to microstresses fulfilling a failure criterion (such as the ideally plastic von Mises criterion calibrated by the uniaxial strength of pure titanium herein), can be assigned to corresponding macroscopic stress states, defining a 'macroscopic', homogenized (ideally plastic) yield criterion of the following, elliptical form:

$$\mathfrak{F}(\Sigma_m, \Sigma_d, p) = \frac{3\varphi}{4(1-\varphi)^2}(\Sigma_m + p)^2 + \frac{1+(2/3)\varphi}{(1-\varphi)^2}\Sigma_d^2 - \frac{f_y^2}{3} = 0 \tag{E.21}$$

with Σ_m and Σ_d as the mean and the equivalent macroscopic stress, reading as

$$\Sigma_m = \frac{1}{3}\text{tr}\boldsymbol{\Sigma} \tag{E.22}$$

and

$$\Sigma_d = \sqrt{\frac{1}{2}\boldsymbol{\Sigma}_d : \boldsymbol{\Sigma}_d}, \quad \boldsymbol{\Sigma}_d = \boldsymbol{\Sigma} - \frac{1}{3}\text{tr}\boldsymbol{\Sigma}\,\mathbf{1} \tag{E.23}$$

and p as the pressure acting inside the pores. It is important to note that p is a state variable independent of $\boldsymbol{\Sigma}$. In particular, p is not equal to hydrostatic part of the macroscopic stress, $1/3$ tr $\boldsymbol{\Sigma}$, as it is sometimes used in the open literature.

For validation of the micromechanics model through our experimental data, we consider a Cartesian base frame with base vectors $\boldsymbol{e}_1$, $\boldsymbol{e}_2$ and $\boldsymbol{e}_3$, where the third axis coincides with the long axis of the cylindrical samples. We consider model predictions for the yield stress in:

1. uniaxial compression without internal pore pressure:

$$\boldsymbol{\Sigma} = \Sigma_{33}\boldsymbol{e}_3 \otimes \boldsymbol{e}_3,$$

$$p = 0,$$

and in

2. triaxial (not hydrostatic) compression with internal pore pressure:

$$\boldsymbol{\Sigma} = -p_0\boldsymbol{e}_1 \otimes \boldsymbol{e}_1 - p_0\boldsymbol{e}_2 \otimes \boldsymbol{e}_2 + \Sigma_{33}\boldsymbol{e}_3 \otimes \boldsymbol{e}_3,$$

$$p = p_0,\ p_0 = 14.5\,\text{MPa},$$

where Σ_{33} is the normal stress related to the axial compression load imposed by the electromechanical machine onto the specimen, irrespective of the pore pressure p_0.

The aforementioned model predictions compare quite well to corresponding, experimentally obtained values (Figure E.10). Consideration of two differently sized porosities in a multistep-homogenization procedure, instead of only one as done herein, might improve the model predictions.

E.6 Conclusions

Triaxial mechanical tests and ultrasound experiments were performed on porous titanium samples of different porosity, in order to determine their Young's moduli and Poisson's ratios, as well as their plastic behavior and yield stresses. The investigations indicate that porous titanium material has a hardening plasticity behavior as seen in load-displacement curves (Figure E.4). Experiments show that yield stress and Young's modulus decrease at increasing porosity (see data points in Figures E.9 and E.10). The experimental results were consistent with poro-micromechanical model predictions based on the stiffness and strength properties of pure titanium, as well as on the sample specific porosity. In addition, the corresponding Mori-Tanaka model for upscaling of elasticity shows that the overall Young's modulus of the porous titanium samples depend nonlinearly and convexly on the porosity (Figure E.9); while a nonlinear homogenization scheme based on effective microstrains in the solid material matrix, shows that the uniaxial yield stress depends more linearly on the porosity and that internal oil pressure increases the yield stress (Figure E.10). However, as the employed oil pressure is by far smaller than the uniaxial yield stress, the aforementioned increase is very small in the present case. This is probably the reason why it could not be clearly confirmed by the experiments. This leads the way to our next step in the described research project, devoted to application of the same oil pressure to materials characterized by a higher porosity, and to application of by far higher oil pressures to materials such as the ones described herein. In addition, we plan an extension of the experimental program towards cyclic loading. This loading condition is highly relevant for the day-to-day use of implants (Hosoda et al. 2006), and also plays an important mechanobiological role (Mizuno et al. 2004).

Acknowledgements

This work was supported in part by the EU Network of Excellence project Knowledge-based Multicomponent Materials for Durable and Safe Performance (KMM-NoE) under the contract no. NMP3-CT-2004-502243. The authors are grateful for helpful discussions with Zbigniew Pakiela, Warsaw University of Technology.

Publication F

Micromechanics of bioresorbable porous CEL2 glass ceramic scaffolds for bone tissue engineering (Malasoma et al. 2008)

Authored by Andrea Malasoma, Andreas Fritsch, Christoph Kohlhauser, Tomasz Brynk, Chiara Vitale-Brovarone, Zbigniew Pakiela, Josef Eberhardsteiner, and Christian Hellmich
Published in *Advances in Applied Ceramics*, Volume 107, pages 277–286

Owing to their stimulating effects on bone cells, ceramics are identified as expressly promising materials for fabrication of tissue engineering (TE) scaffolds. To ensure the mechanical competence of TE scaffolds, it is of central importance to understand the impact of pore shape and volume on the mechanical behaviour of the scaffolds, also under complex loading states. Therefore, the theory of continuum micromechanics is used as basis for a material model predicting relationships between porosity and elastic/strength properties. The model, which mathematically expresses the mechanical behaviour of a ceramic matrix (based on a glass system of the type SiO_2-P_2O_5-CaO-MgO-Na_2O-K_2O; called CEL2) in which interconnected pores are embedded, is carefully validated through a wealth of independent experimental data. The remarkably good agreement between porosity based model predictions for the elastic and strength properties of CEL2-based porous scaffolds and corresponding experimentally determined mechanical properties underlines the great potential of micromechanical modelling for speeding up the biomaterial and tissue engineering scaffold development process – by delivering reasonable estimates for thematerial behaviour, also beyond experimentally observed situations.

Notation

$\mathbb{A}_r$	fourth order strain concentration tensor of phase r
$\mathbb{A}_S$	fourth order strain concentration tensor of solid phase (dense CEL2 glass ceramic)
$\mathbb{A}_{por}$	fourth order strain concentration tensor of pores
a	typical cross-sectional dimension of a CEL2-based porous biomaterial sample
$\mathbb{C}_{hom}$	fourth order homogenised stiffness tensor
C_{ijkl}	components of fourth order homogenised stiffness tensor
$\mathbb{C}_{por}$	fourth order stiffness tensor of pores
$\mathbb{C}_S$	fourth order stiffness tensor of solid phase (dense CEL2 glass ceramic)
d	characteristic length of inhomogeneity within an RVE
$\boldsymbol{E}$	second order 'macroscopic' strain tensor
$\boldsymbol{E}_d$	deviatoric part of macroscopic strain tensor
E_S	Young's modulus of solid phase (dense CEL2 glass ceramic)
E_{exp}	experimentally determined Young's modulus of porous CEL2-based biomaterial
$\bar{E}_{exp}$	mean over all experimentally determined Young's moduli of porous CEL2-based biomaterial
E_{hom}	homogenised Young's modulus of porous CEL2-based biomaterial
$\bar{e}$	mean of relative error between predictions and experiments
e_S	standard deviation of relative error between predictions and experiments
$\underline{e}_1$	unit base vector of Cartesian reference base frame
f	ultrasonic excitation frequency
$\mathfrak{f}(\boldsymbol{\sigma}) = 0$	boundary of elastic domain of solid material phase, in space of microstresses
$\mathfrak{F}(\boldsymbol{\Sigma}) = 0$	boundary of elastic domain of porous CEL2-based biomaterial, in space of macrostresses
g_1, g_2	functions for determination of homogenised elastic constants k_{hom} and μ_{hom} [see Eq. (F.18)]
$\mathbb{I}$	fourth-order identity tensor
$\mathbb{J}$	volumetric part of fourth-order identity tensor $\mathbb{I}$
$\mathbb{K}$	deviatoric part of fourth-order identity tensor $\mathbb{I}$
k_{DS}^j, k_{DS}^{j+1}	homogenised bulk moduli of step j and $j+1$ in a Differential Scheme
k_S	Bulk modulus of solid phase (dense CEL2 glass ceramic)
k_{hom}	homogenised bulk modulus of porous CEL2-based biomaterial
L	characteristic length of a structure containing an RVE
ℓ_{RVE}	characteristic length of RVE of porous CEL2-based biomaterial
l	length of ultrasonic path
M	mass of a porous CEL2-based biomaterial sample

RVE	representative volume element
r	index for phases
$\mathbb{S}_{sph}$	fourth order Eshelby tensor for spherical inclusion embedded in isotropic matrix with stiffness $\mathbb{C}_S$
t	transition time of an ultrasonic wave through a CEL2-based biomaterial sample
tr	trace of a second order tensor
V	volume of a porous CEL2-based biomaterial sample
V_{por}	volume of pores within an RVE of porous CEL2-based biomaterial
V_S	volume of the solid phase (dense CEL2 glass ceramic) within an RVE of porous CEL2-based biomaterial
V_{RVE}	volume of an RVE of porous CEL2-based biomaterial
v	propagation velocity of ultrasonic wave within a CEL2-based biomaterial sample
$\underline{x}$	position vector within an RVE
$\Delta\varphi$	pore increment in a Differential Scheme
Δx	very small volume fraction of homogenised material in a Differential Scheme, to be replaced by pores
δ_{ij}	Kronecker delta
$\boldsymbol{\varepsilon}$	second order microscopic strain tensor
$\boldsymbol{\varepsilon}_d$	deviatoric part of microscopic strain tensor
ε_d	equivalent deviatoric microscopic strain
ε_d^{eff}	effective deviatoric microscopic strain
$\boldsymbol{\varepsilon}_{por}$	average microscopic strain in pore phase
$\boldsymbol{\varepsilon}_r$	average microscopic strain in phase r
$\boldsymbol{\varepsilon}_S$	average microscopic strain in solid phase (dense CEL2 glass ceramic)
λ	ultrasonic wave length
μ_{DS}^{j}, μ_{DS}^{j+1}	homogenised shear moduli of step j and $j+1$ in a Differential Scheme
μ_{hom}	homogenised shear modulus of porous CEL2-based biomaterial sample
ν_S	Poisson's ratio of solid phase (dense CEL2 glass ceramic)
ν_{hom}	homogenised Poisson's ratio of porous CEL2-based biomaterial sample
$\underline{\xi}$	displacements within an RVE and at its boundary
ρ	material mass density of porous CEL2-based biomaterial sample
ρ_S	material mass density of solid phase (dense CEL2 glass ceramic)
$\boldsymbol{\Sigma}$	second order 'macroscopic' stress tensor

$\boldsymbol{\Sigma}_d$	deviatoric part of macroscopic stress tensor
$\Sigma^{ult,c}_{pred}$	model predicted uniaxial compressive strength of porous CEL2-based biomaterial
$\Sigma^{ult,c}_{exp}$	experimentally determined uniaxial compressive strength of porous CEL2-based biomaterial
$\boldsymbol{\sigma}$	second order 'microscopic' stress tensor
$\boldsymbol{\sigma}_d$	deviatoric part of microscopic stress tensor
σ_d	equivalent deviatoric microscopic stress
σ^{eff}_d	effective deviatoric microscopic stress
$\boldsymbol{\sigma}_S$	average microscopic stress in solid phase (dense CEL2 glass ceramic)
τ^{ult}_S	shear strength of dense CEL2 glass ceramic
τ^{ult}	shear strength
φ	volume fraction of pores within an RVE of porous CEL2-based biomaterial
∂V	boundary of an RVE
$\mathbf{1}$	second order identity tensor
$\langle(.)\rangle_V = 1/V \int_V (.) \mathrm{d}V$	average of quantity (.) over volume V
$\cdot$	first order tensor contraction
$:$	second order tensor contraction
$\otimes$	dyadic product of tensors

F.1 Introduction

Bone replacements are needed for many orthopaedic, maxillofacial and craniofacial surgeries. The latter may be required due to e.g. trauma or bone neoplasia. Hence, bone regeneration is an increasingly important clinical need. Autografts, allografts and xenografts can be used as bone substitutes; autografts are still considered as the best choice, because of their ability to support osteoinduction and osteogenesis, but considerable drawbacks are associated with the need for further surgery and with donor site morbidity. Allografts and xenografts represent a promising alternative, but they show worse bone induction properties, lower integration rates and non-negligible risks of viral contamination. For these reasons, artificial grafts (also called scaffolds) are interesting candidates to stimulate bone regeneration.

The term scaffold refers to a structure, realised with natural or synthesised materials, which is able to promote cellular regeneration and to guide bone regeneration. Therefore, synthetic scaffolds may be seeded with carefully chosen biological cells and/or growth factors: this is referred to as tissue engineering (Langer and Vacanti 1993). Within this concept, the main role of a scaffold is to assure a mechanical support to the growing tissue, to guide this growth and to induce correct development of the bony organ. Due to their stimulating effects on bone cells, ceramics (such as hydroxyapatite (Akao et al. 1981; Verma et al. 2006), β-tricalcium phosphate (Charrière et al. 2001), bioactive glasses (Hench and Jones 2005; Boccaccini et al.

2005), or glass ceramics (Vitale-Brovarone et al. 2007)) are identified as expressly promising materials for fabrication of tissue engineering scaffolds.

However, the design of such scaffolds is still a great challenge since (at least) two competing requirements must be fulfilled:

1. on the one hand, the scaffold must exhibit a sufficient mechanical competence, i.e. stiffness and strength comparable to natural bones;

2. on the other hand, once the scaffold would be implanted into the living organism, it should be continuously resorbed and replaced by natural bones. This typically requires a sufficient pore space (pore size in the range of hundred micrometres and porosity of more than 50-60% (Cancedda et al. 2007)), which discriminates the aforementioned mechanical properties, and therefore competes with the first requirement.

For finding a good balance between these competing requirements, it is of central importance to understand the impact of pore shape and volume on the mechanical behaviour of the scaffolds, also under complex loading states. In order to contribute to this understanding, the authors started a multidisciplinary activity driven forward by physicists, chemists, material scientists, and engineering mechanicians. While the authors' endeavours comprised state of the art processing and characterisation techniques, ranging all the way from microscopy to mechanical and acoustical testing, the focus of the present contribution is on an engineering science based synthesis tool for consistent explanation of the experimental data: in more detail, the theory of continuum micromechanics (Suquet 1997a; Zaoui 2002) provides the authors with the basis for a material model predicting relationships between porosity and elastic/strength properties. The model, which mathematically expresses the mechanical behaviour of a ceramic matrix in which interconnected pores are embedded (see Section F.3), is carefully validated through a wealth of independent experimental data (see Section F.4). The latter are gained from geometrical and weighing measurements and from mechanical tests on CEL2 biomaterials (see Section F.4). These biomaterials are based on a glass system of the type SiO_2-P_2O_5-CaO-MgO-Na_2O-K_2O, the production and microstructural morphology of which will be given in Section F.2. The remarkably good agreement between porosity based model predictions for elastic and strength properties of CEL2-based porous scaffolds and corresponding experimentally determined mechanical properties (see Section F.4) underlines the great potential of micromechanical modelling for speeding up the biomaterial and tissue engineering scaffold development process – by delivering reasonable estimates for the material behaviour, also beyond experimentally observed situations. A related discussion concludes the present paper (see Section F.5).

F.2 Processing and microstructural characterisation of CEL2 biomaterials before and after bioactivity treatment

The production of glass ceramic tissue engineering scaffolds with different porosities was based on a glass called CEL2 (Vitale-Brovarone et al. 2007). This glass belongs to the system SiO_2-P_2O_5-CaO-MgO-Na_2O-K_2O, with the following molar composition: 45% SiO_2, 3% P_2O_5, 26% CaO, 7% MgO, 15% Na_2O, 4% K_2O. CEL2 was prepared by melting the raw products in a platinum crucible at 1400°C for 1 h and by quenching the melt in cold water to obtain a frit that was finally ground and sieved. This resulted in a final grain size of less than 30 μm.

The porous scaffolds were produced by means of two different methods:

1. the replication technique based on a polymeric sponge
2. the burning-out method based on a mixture of glass and organic powders.

In the latter method, different quantities of an (polyethylene) organic powder with grain sizes of 100-600 μm are mixed with the aforementioned CEL2 powder, leading to different porosities of the end product. Subsequently, the mixture is pressed, then it passes through a heat treatment where the polymer burns, leaving pores on the substrate; finally, the powders are sintered. As an alternative production technique, the replication method involves the impregnation of a polymeric template with a suitable powder suspension (slurry). The chosen template possesses a porous microstructure and, after the impregnation phase, the template undergoes a thermal treatment that burns out the organic phase and sinters the inorganic one.

To check the bioactivity requirement given in Section F.1, some of the replication technique based 3D scaffolds were treated in simulated body fluid (SBF) for one week (sample 'B' in Tables F.2 and F.3) and for four weeks (sample 'D' in Tables F.2 and F.3) respectively, in order to study the formation of hydroxyapatite crystals on the sintered struts (Figure F.2). In addition, 3D scaffolds were also soaked in a buffered medium, trishydroxymethylaminomethane (standardly abbreviated as tris), again for one week (sample 'C' in Tables F.2 and F.3) and four weeks (sample 'D' in Tables F.2 and F.3) respectively, so as to assess the scaffolds bioresorption with time.

The microstructural morphology of the scaffolds was studied by means of scanning electron microscopy (SEM). The replication technique allows for realisation of strut like morphologies inspired by trabecular bone architecture [Figure F.1(a)-(b)], while the powder mixture technique results in porous matrix type morphologies [Figure F.1(c)-(d)]. In both cases, the pore sizes related to the tailored (macro) porosity range between 100 and 500 μm. Moreover, the sintering process induces a microporosity (with characteristic length of 15 μm) important for adhesion of proteins and cells. After soaking in SBF or tris at 37°C, a new phase formed on

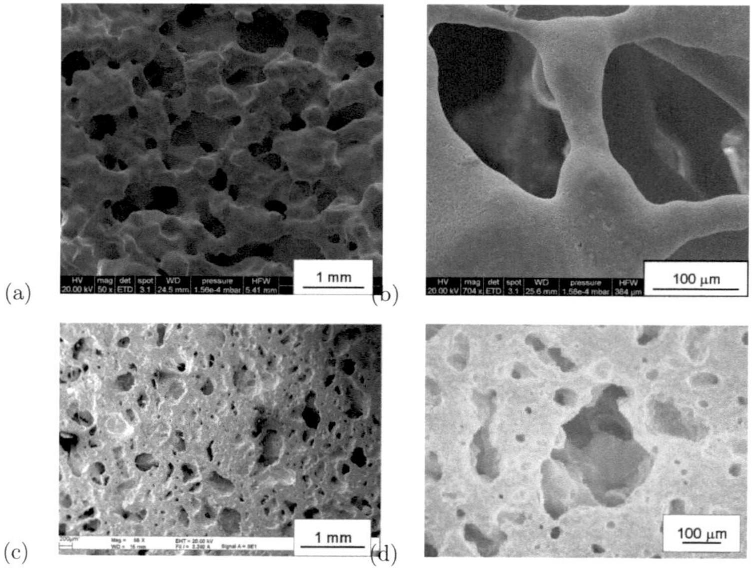

Figure F.1: Scanning electron micrographs of CEL2 glass ceramic scaffolds at different resolutions, produced by replication method (a)-(b), and by burning-out method (c)-(d)

the pore surfaces (Figure F.2), showing the remarkable bioactivity of the material. In SBF, the chemical composition of this new phase was confirmed to be close to hydroxyapatite, by means of X-ray diffraction (XRD) and energy dispersion spectrometry (EDS). The pH variations in the pores during the soaking of the scaffolds were also monitored: ranging between 7.4 and 8, they fall into the moderately alkaline conditions preferred by the osteoblasts, the biological cells building up an extracellular bone matrix.

Next, the microstructural information contained in Figures F.1 and F.2 is reduced to the features which are essential to capture the mechanical behaviour of the scaffolds. Therefore, the authors will not distinguish between the solid glass ceramic substance and the new phase initiated through treatment in SBF or tris. The relevance of this simplification will be underlined in the section devoted to model validation. The model itself will be cast in the framework of continuum micromechanics, as is detailed next.

Figure F.2: Scanning electron micrograph of CEL2 glass ceramic scaffold after one week of soaking in SBF

F.3 Micromechanical model

F.3.1 Fundamentals of continuum micromechanics – representative volume element

In continuum micromechanics (Suquet 1997a; Zaoui 2002; Hill 1963) a material is understood as a macrohomogeneous, but microheterogeneous body filling a representative volume element (RVE) with characteristic length ℓ_{RVE}, $\ell_{RVE} \gg d$, d standing for the characteristic length of inhomogeneities within the RVE, and $\ell_{RVE} \ll L$, L standing for the characteristic lengths of the geometry or loading of a structure built up by the material defined on the RVE. In general, the microstructure within each RVE is so complicated that it cannot be described in complete detail. Therefore, quasihomogeneous subdomains with known physical quantities (such as volume fractions, elastic or strength properties) are reasonably chosen. They are called material phases. The 'homogenised' mechanical behaviour of the overall material, i.e. the relation between homogeneous deformations acting on the boundary of the RVE and resulting (average) stresses, or the ultimate stresses sustainable by the RVE, can then be estimated from the mechanical behaviour of the aforementioned homogeneous phases (representing the inhomogeneities within the RVE), their dosages within the RVE, their characteristic shapes, and their interactions.

F.3.2 Micromechanical representation of CEL2-based biomaterial

An RVE of CEL2-based biomaterial is considered, with characteristic length ℓ_{RVE}=55 mm and with volume V_{RVE}, hosting spherical, empty pores with characteristic size d=100-500 μm

$\ll \ell_{RVE}$, with volume V_{por} and volume fraction φ ($=V_{por}/V_{RVE}$). These pores are embedded in a solid matrix with volume V_S and volume fraction (1-φ) (see Figure F.3).

Homogeneous ('macroscopic') strains $\boldsymbol{E}$ are imposed onto the RVE, in terms of displacements $\underline{\xi}$ at its boundary ∂V

$$\forall \underline{x} \in \partial V: \quad \underline{\xi}(\underline{x}) = \boldsymbol{E} \cdot \underline{x} \tag{F.1}$$

with $\underline{x}$ as the position vector within the RVE. As a consequence, the resulting kinematically compatible microstrains $\boldsymbol{\varepsilon}(\underline{x})$ throughout the RVE with volume V_{RVE} fulfil the average condition (Hashin 1983)

$$\boldsymbol{E} = \langle \boldsymbol{\varepsilon} \rangle = \frac{1}{V_{RVE}} \int_{V_{RVE}} \boldsymbol{\varepsilon}(\underline{x})\, \mathrm{d}V = (1-\varphi)\, \boldsymbol{\varepsilon}_S + \varphi\, \boldsymbol{\varepsilon}_{por} \tag{F.2}$$

with

$$\boldsymbol{\varepsilon}_S = \frac{1}{V_S} \int_{V_S} \boldsymbol{\varepsilon}(\underline{x})\, \mathrm{d}V, \quad \boldsymbol{\varepsilon}_{por} = \frac{1}{V_{por}} \int_{V_{por}} \boldsymbol{\varepsilon}(\underline{x})\, \mathrm{d}V, \quad V_S + V_{por} = V_{RVE} \tag{F.3}$$

Equation (F.2) provides a link between 'micro' and 'macro' strains. Thereby, $\boldsymbol{\varepsilon}_S$ and $\boldsymbol{\varepsilon}_{por}$ are the averages of the (micro)strain tensor fields, over the solid and the porous phase respectively [see equation (F.3)]. Analogously, homogenised ('macroscopic') stresses $\boldsymbol{\Sigma}$ are defined as the spatial average over the RVE of the microstresses $\boldsymbol{\sigma}(\underline{x})$

$$\boldsymbol{\Sigma} = \langle \boldsymbol{\sigma} \rangle = \frac{1}{V_{RVE}} \int_{V_{RVE}} \boldsymbol{\sigma}\, \mathrm{d}V = (1-\varphi)\, \boldsymbol{\sigma}_S \tag{F.4}$$

with $\boldsymbol{\sigma}_S$ as the average of the (micro)stress tensor field over the solid phase.

F.3.3 Constitutive behaviour of CEL2 and pores

The solid phase (consisting of dense CEL2 glass ceramic, and in case of samples tested for biocompatibility, also of tris or SBF-derived substances) inside the RVE V_{RVE} behaves linear elastically

$$\boldsymbol{\sigma} = \mathbb{C}_S : \boldsymbol{\varepsilon}_S \tag{F.5}$$

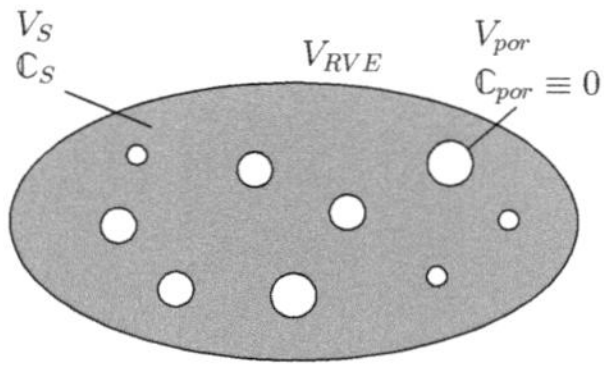

Figure F.3: Micromechanical representation of CEL2-based biomaterial: macropores of porosity φ are embedded and interconnected within dense (microporous) solid glass substance with elasticity tensor $\mathbb{C}_S$

with $\mathbb{C}_S = 3k_S\mathbb{J} + 2\mu_S\mathbb{K}$ as the isotropic elastic stiffness of the solid phase; with bulk modulus k_S and shear modulus μ_S. $\mathbb{J} = 1/3\mathbf{1} : \mathbf{1}$ and $\mathbb{K} = \mathbb{I} - \mathbb{J}$ are the volumetric and the deviatoric part of the fourth order identity tensor $\mathbb{I}$, with components $I_{ijkl} = 1/2(\delta_{ik}\delta_{jl} + \delta_{il}\delta_{kj})$; the components of the second order unit tensor $\mathbf{1}$, δ_{ij} (Kronecker delta), read as $\delta_{ij}=1$ for $i=j$ and $\delta_{ij}=0$ for $i\neq j$. The pores are empty, therefore $\mathbb{C}_{por} = \mathbb{O}$. The load bearing capacity of the solid phase is bounded according to a von Mises-type failure criterion, reading as

$$\mathfrak{f}(\boldsymbol{\sigma}(\underline{x})) = \sigma_d(\underline{x}) - \tau^{ult} = 0 \tag{F.6}$$

where τ^{ult} is the shear strength of the solid phase, and σ_d is the equivalent deviatoric microscopic stress, reading as

$$\sigma_d(\underline{x}) = \sqrt{\frac{1}{2}\boldsymbol{\sigma}_d(\underline{x}) : \boldsymbol{\sigma}_d(\underline{x})} \tag{F.7}$$

with

$$\boldsymbol{\sigma}_d(\underline{x}) = \boldsymbol{\sigma}(\underline{x}) - \frac{1}{3}\operatorname{tr}\boldsymbol{\sigma}(\underline{x})\mathbf{1} \tag{F.8}$$

as the deviatoric part of the microscopic stress tensor $\boldsymbol{\sigma}$.

F.3.4 Homogenisation of elastic properties

Homogenised ('macroscopic') stresses and strains, $\boldsymbol{\Sigma}$ and $\boldsymbol{E}$, are related by the homogenised ('macroscopic') stiffness tensor $\mathbb{C}_{hom}$

$$\boldsymbol{\Sigma} = \mathbb{C}_{hom} : \boldsymbol{E} \tag{F.9}$$

which needs to be linked to the solid stiffness $\mathbb{C}_S$, as well as to the shape, and to the spatial arrangement of the phases (solid glass ceramic substance and pores). This link is based on the linear relation between the homogenised ('macroscopic') strain $\boldsymbol{E}$ and the average ('microscopic') strain $\boldsymbol{\varepsilon}_r$, resulting from the superposition principle valid for linear elasticity [equation (F.5)] (Hill 1963). This relation is expressed in terms of the fourth order concentration tensors $\mathbb{A}_r$ of each of the phases r (r=S or por)

$$\boldsymbol{\varepsilon}_r = \mathbb{A}_r : \boldsymbol{E} \tag{F.10}$$

which implies, together with equation (F.2), that

$$(1-\varphi)\,\mathbb{A}_S + \varphi\,\mathbb{A}_{por} = \mathbb{I} \tag{F.11}$$

Insertion of equation (F.10) into equation (F.5) and averaging over all phases according to equation (F.4) leads to

$$\boldsymbol{\Sigma} = (1-\varphi)\,\mathbb{C}_S : \mathbb{A}_S : \boldsymbol{E} \tag{F.12}$$

From equations (F.12) and (F.9), the sought relation between the phase stiffness tensor $\mathbb{C}_S$ and the overall homogenised stiffness $\mathbb{C}_{hom}$ of the RVE can be identified

$$\mathbb{C}_{hom} = (1-\varphi) : \mathbb{C}_S : \mathbb{A}_S = \mathbb{C}_S : (\mathbb{I} - \varphi\,\mathbb{A}_{por}) \tag{F.13}$$

If the porosity is very small, $\varphi \ll 1$ (dilute dispersion of pores), the mechanical interactions between the pores can be neglected. In this case, the macroscopic strains $\boldsymbol{E}$ acting on the RVE of Figure F.3 can be set equal to those acting on the remote boundary of an infinite matrix made up by the solid phase, a matrix which hosts one pore like inclusion. Under this condition, the homogeneous (microscopic) strains $\boldsymbol{\varepsilon}_{por}$ within a spherical empty pore follows from Eshelby's 1957 problem (Eshelby 1957), and read as

$$\boldsymbol{\varepsilon}_{por} = \underbrace{[\mathbb{I} - \mathbb{S}_{sph}]^{-1}}_{\mathbb{A}_{por}} : \boldsymbol{E} \tag{F.14}$$

whereby $\mathbb{A}_{por}$ follows from equation (F.10). The fourth order Eshelby tensor $\mathbb{S}_{sph}$ accounts for the morphology of the inclusion. For spheres, it reads as

$$\mathbb{S}_{sph} = \frac{3k_S}{3k_S + 4\mu_S}\mathbb{J} + \frac{6(k_S + 2\mu_S)}{5(3k_S + 4\mu_S)}\mathbb{K} \tag{F.15}$$

Use of equations (F.14) and (F.15) in equation (F.13) yields the so called 'dilute estimate' for the stiffness of a porous material with spherical pores. In the present situation, however, this estimate needs to be extended to the case of higher porosities made up by interconnected pores (see Figures F.1 and F.2). Therefore, the so called Differential Scheme is used (Boucher 1976; McLaughlin 1977; Molinari and El Mouden 1996; Dormieux and Lemarchand 2001). Initially, a very small volume fraction of pores $\Delta\varphi$ is introduced into the solid matrix and the material is homogenised via equations (F.14), (F.15) and (F.13). The following steps consist in (i) removing a very small portion $\Delta x \ll 1$ of the previously homogenised material (containing already some porosity), in (ii) replacing it by the same volume fraction of pores (see Figure F.4), and in (iii) homogenisation of the slightly more porous material. Thereby, the overall porosity increases by the increment $\Delta\varphi$.

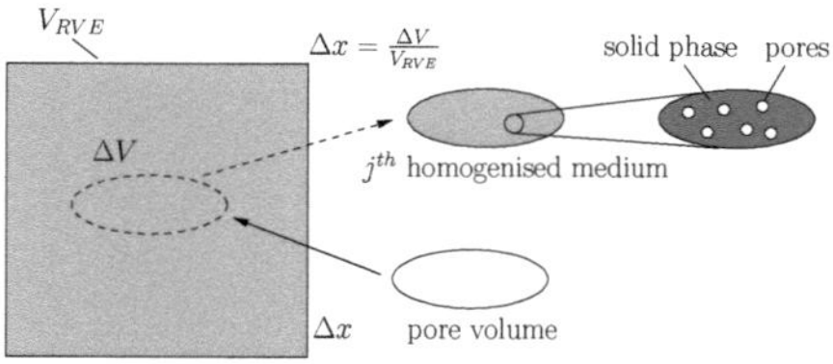

Figure F.4: Schematical representation of Differential Scheme, in the line of (Dormieux and Lemarchand 2001)

$$\Delta\varphi = -\varphi\Delta x + \Delta x = (1-\varphi)\Delta x, \quad \Delta x \ll 1 \tag{F.16}$$

Repeating this removal and introduction of small volume fractions, followed by subsequent homogenisation, leads to an iteration scheme of the form (Dormieux et al. 2006b)

$$\begin{aligned} k_{DS}^{j+1} &= k_{DS}^{j}\left[1-\left(1+\frac{3k_{DS}^{j}}{4\mu_{DS}^{j}}\right)\Delta x\right] \\ \mu_{DS}^{j+1} &= \mu_{DS}^{j}\left[1-\left(5\frac{3k_{DS}^{j}+4\mu_{DS}^{j}}{9k_{DS}^{j}+8\mu_{DS}^{j}}\right)\Delta x\right] \end{aligned} \quad \text{(F.17)}$$

with k_{DS}^{j} and μ_{DS}^{j} as the homogenised moduli after the jth homogenisation step. Realising scheme (F.17) for the limit case $\Delta\varphi \to 0$, as long as the actual porosity is reached, $\sum_j \varphi_j = \varphi$, yields the differential estimate (Dormieux et al. 2006b)

$$\begin{aligned} g1 &= \frac{(1+4\mu_S/3k_S)(\mu_{hom}/\mu_S)^3}{2-(1-4\mu_S/3k_S)(\mu_{hom}/\mu_S)^{3/5}} - (1-\varphi)^6 = 0 \\ g2 &= \frac{\mu_{hom}}{\mu_S} - \frac{(1-4/3\,\mu_{hom}/k_{hom})^{5/3}}{(1-4/3\,\mu_S/k_S)^{5/3}} = 0 \end{aligned} \quad \text{(F.18)}$$

with k_{hom} and μ_{hom} as the bulk and the shear modulus of the homogenised stiffness tensor $\mathbb{C}_{hom}$, $\mathbb{C}_{hom} = 3k_{hom}\mathbb{J} + 2\mu_{hom}\mathbb{K}$. Equation (F.18) is valid as long as Poisson's ratio $\nu_S = (3k_S - 2\mu_S)/(6k_S + 2\mu_S)$ is larger than 0.2 (see (Dormieux et al. 2006b)). Finally, standard isotropic elasticity relates k_{hom} and μ_{hom} to the Young's modulus E_{hom}

$$E_{hom} = \frac{9k_{hom}\mu_{hom}}{3k_{hom} + \mu_{hom}} \quad \text{(F.19)}$$

F.3.5 Upscaling of failure properties

In order to determine the effective failure properties resulting from local failure characteristics [equation (F.17)], we are left with relating the local strains and stresses to corresponding macroscopic quantities. In contrast to the homogenised elastic properties, which can be derived from (first order) averages of microstrains and microstresses over the material phases [see $\boldsymbol{\varepsilon}_S$ and $\boldsymbol{\varepsilon}_{por}$ in equation (F.3)], homogenisation of strength properties calls for additional information on the heterogeneity of these microquantities, i.e. the strain or stress peaks inside the microstructure (possibly cancelled out through averaging) need to be appropriately considered. This heterogeneity can reasonably be considered through so called effective microstrains ε_d^{eff} (Kreher 1990; Dormieux et al. 2002; Barthélémy and Dormieux 2003, 2004) (see (Fritsch et al. 2007a,b, 2009a) for application to hydroxyapatite ceramics), such as the square root of the average over the solid material phase, of the squares of the equivalent deviatoric (micro) strains $\varepsilon_d(\underline{x})$,

$$\varepsilon_d^{eff} = \sqrt{\frac{1}{V_S}\int_{V_S} \varepsilon_d^2(\underline{x})\mathrm{d}V} \quad \text{(F.20)}$$

with

$$\varepsilon_d(\underline{x}) = \sqrt{\frac{1}{2}\boldsymbol{\varepsilon}_d(\underline{x}) : \boldsymbol{\varepsilon}_d(\underline{x})} \tag{F.21}$$

with the deviatoric microstrain tensor

$$\boldsymbol{\varepsilon}_d(\underline{x}) = \boldsymbol{\varepsilon}(\underline{x}) - \frac{1}{3}\,\mathrm{tr}\,\boldsymbol{\varepsilon}(\underline{x})\,\mathbf{1} \tag{F.22}$$

and with tr $\boldsymbol{\varepsilon}$ as the trace of the microscopic strain tensor. Energy considerations (Dormieux et al. 2002) allow for determination of the effective deviatoric strain ε_d^{eff} from the macroscopic strains $\boldsymbol{E}$, according to

$$\varepsilon_d^{eff,2} = \frac{1}{2(1-\varphi)}\left[\frac{1}{2}\frac{\partial k_{hom}}{\partial \mu_S}(\mathrm{tr}\,\boldsymbol{E})^2 + \frac{\partial \mu_{hom}}{\partial \mu_S}\boldsymbol{E}_d : \boldsymbol{E}_d\right] \tag{F.23}$$

with tr $\boldsymbol{E}$ and $\boldsymbol{E}_d$ as the trace and the deviatoric part of the macroscopic strain tensor $\boldsymbol{E}$. The definition of $\boldsymbol{E}_d$ is analogous to equation (F.22). The derivations of k_{hom} and μ_{hom} with respect to μ_S are obtained via implicit differentiation of equation (F.18), leading to

$$\frac{\partial \mu_{hom}}{\partial \mu_S} = \frac{-\frac{\partial g_1}{\partial \mu_S}}{\frac{\partial g_1}{\partial \mu_{hom}}}, \quad \frac{\partial k_{hom}}{\partial \mu_S} = \frac{-\left(\frac{\partial g_2}{\partial \mu_{hom}}\frac{\partial \mu_{hom}}{\partial \mu_S} + \frac{\partial g_2}{\partial \mu_S}\right)}{\frac{\partial g_2}{\partial k_{hom}}} \tag{F.24}$$

whereby

$$\begin{aligned}
\frac{\partial g_1}{\partial \mu_S} &= -\frac{6\mu_{hom}^3(9k_S + 8\mu_S)\left(4\mu_{hom}\mu_S + 5\mu_S\left(\frac{\mu_{hom}}{\mu_S}\right)^{2/5}k_S - 2\mu_{hom}k_S\right)}{\mu_S\mathcal{N}}, \\
\frac{\partial g_1}{\partial \mu_{hom}} &= \frac{6\mu_{hom}^2(3k_S + 4\mu_S)\left(8\mu_{hom}\mu_S + 15\mu_S\left(\frac{\mu_{hom}}{\mu_S}\right)^{2/5}k_S - 6\mu_{hom}k_S\right)}{\mathcal{N}}, \\
\mathcal{N} &= 5\mu_S^4\left(6k_S - 3\left(\frac{\mu_{hom}}{\mu_S}\right)^{3/5}k_S + 4\left(\frac{\mu_{hom}}{\mu_S}\right)^{3/5}\mu_S\right)^2\left(\frac{\mu_{hom}}{\mu_S}\right)^{2/5}, \\
\frac{\partial g_2}{\partial \mu_S} &= -\frac{\mu_{hom}}{\mu_S^2} - \frac{20\left(1-\frac{4\mu_{hom}}{3k_{hom}}\right)^{5/3}}{9k_S\left(1-\frac{4\mu_S}{3k_S}\right)^{8/3}}, \quad \frac{\partial g_2}{\partial \mu_{hom}} = \frac{1}{\mu_S} + \frac{20\left(1-\frac{4\mu_{hom}}{3k_{hom}}\right)^{2/3}}{9k_{hom}\left(1-\frac{4\mu_S}{3k_S}\right)^{5/3}}, \\
\frac{\partial g_2}{\partial k_{hom}} &= \frac{20\mu_{hom}\left(1-\frac{4\mu_{hom}}{3k_{hom}}\right)^{2/3}}{9k_{hom}^2\left(1-\frac{4\mu_S}{3k_S}\right)^{5/3}}
\end{aligned} \tag{F.25}$$

The macroscopic strains $\boldsymbol{E}$ and $\boldsymbol{E}_d$ in equation (F.23) are related to the corresponding macroscopic stress states via the homogenised stiffness tensor $\mathbb{C}_{hom}$ [see equation (F.9)]. In equation (F.6), stress peaks of $\sigma_d(\underline{x})$ are left to be estimated by the effective microstress σ_d^{eff}. The latter reads as

$$\sigma_d^{eff} = 2\mu_S\,\varepsilon_d^{eff} \tag{F.26}$$

Insertion of equation (F.26), together with equations (F.18)-(F.25) and (F.9), into the microscopic failure criterion (F.6) with $\sigma_d(\underline{x}) \approx \sigma_d^{eff}$, delivers an elastic limit criterion for macroscopic stress states (representing ultimate strength in the case of brittle materials), as function of the porosity φ

$$\mathfrak{F}(\mathbf{\Sigma}) = \frac{2\mu_S}{\sqrt{2(1-\varphi)}}\left[\frac{1}{2}\frac{\partial k_{hom}}{\partial \mu_S}\left(\frac{\mathrm{tr}\,\mathbf{\Sigma}}{3k_{hom}}\right)^2 + \frac{\partial \mu_{hom}}{\partial \mu_S}\frac{\mathbf{\Sigma}_d : \mathbf{\Sigma}_d}{2\mu_{hom}^2}\right]^{1/2} - \tau^{ult} = 0 \qquad \text{(F.27)}$$

with tr $\mathbf{\Sigma}$ and $\mathbf{\Sigma}_d$ as the trace and the deviatoric part of the macroscopic stress tensor $\mathbf{\Sigma}$. The definition of $\mathbf{\Sigma}_d$ is analogous to equation (F.22).

In particular, strength model (F.27) will be evaluated for stress states related to uniaxial compression $\mathbf{\Sigma} = \Sigma\, \boldsymbol{e}_1 \otimes \boldsymbol{e}_1$, yielding an estimate for the macroscopic uniaxial compressive strength

$$\Sigma_{pred}^{ult,c} = \frac{\left[9(2\nu_{hom}-1)^2\frac{\partial k_{hom}}{\partial \mu_S} + 12(\nu_{hom}+1)^2\frac{\partial \mu_{hom}}{\partial \mu_S}\right]^{1/2}(1-\varphi)^{1/2}E_{hom}\tau^{ult}}{\left[3(2\nu_{hom}-1)^2\frac{\partial k_{hom}}{\partial \mu_S} + 4(2\nu_{hom}+1)^2\frac{\partial \mu_{hom}}{\partial \mu_S}\right]\mu_S} \qquad \text{(F.28)}$$

In equation (F.28), ν_{hom} is Poisson's ratio of the homogenised material

$$\nu_{hom} = \frac{3k_{hom} - 2\mu_{hom}}{6k_{hom} + 2\mu_{hom}} \qquad \text{(F.29)}$$

F.4 Model validation

F.4.1 Strategy for model validation through independent test data

Validation of the micromechanical representation of CEL2-based biomaterials will rest on two independent experimental sets, related to dense CEL2 glass ceramics and to samples of (macro)porous biomaterials: biomaterial specific macroscopic (homogenised) Young's moduli E_{hom} and uniaxial compressive strengths $\Sigma_{pred}^{ult,c}$, predicted by the micromechanics model (see Section F.3) on the basis of biomaterial independent ('universal') elastic and strength properties of pure CEL2-glass (experimental set I, see Section F.4.2) for biomaterial specific porosities φ (experimental set IIa, see Section F.4.3), are compared to corresponding biomaterial specific experimentally determined Young's moduli E_{exp} (experimental set IIb-1, see Section F.4.4) and uniaxial compressive strength values $\Sigma_{exp}^{ult,c}$ (experimental set IIb-2, see Section F.4.5).

F.4.2 'Universal' mechanical properties of dense CEL2 glass ceramics – experimental set I

Acoustic experiments (Kohlhauser et al. 2009) reveal the isotropic elastic constants for dense CEL2 glass ceramic, its Young's modulus $E_S = 85.3$ GPa, and its Poisson's ratio $\nu_S = 0.25$

(equivalent to bulk modulus $k_S = E_S/3/(1-2\nu_S) = 56.9$ GPa and shear modulus $\mu_S = E_S/2/(1+\nu_S) = 34.1$ GPa (see also Table F.1). The authors are not aware of reliable direct strength tests on dense CEL2 glass ceramics. However, ceramic biomaterials made of hydroxyapatite with a microporosity similar to that of the herein investigated materials exhibit a typical shear strength of $\tau^{ult} = 9.8$ MPa (Charrière et al. 2001), which will be considered as representative for dense (microporous) CEL2 glass ceramic (Table F.1).

Young's modulus E_S	85.3 GPa	from (Kohlhauser et al. 2009)
Poisson's ratio ν_S	0.25	from (Kohlhauser et al. 2009)
Shear strength τ_S^{ult}	9.8 MPa	from (Charrière et al. 2001)

Table F.1: 'Universal' (biomaterial-independent) isotropic phase properties of dense CEL2 glass ceramic (=solid phase in Figure F.3)

F.4.3 Sample specific porosities of CEL2-based biomaterials – experimental set IIa

The porosity of the investigated CEL2-based samples was determined from measurements of their masses M and volumes V, according to

$$\varphi = 1 - \frac{M}{V\,\rho_S} \tag{F.30}$$

whereby $\rho_S = 2.6$ g/cm^3 is the mass density of the dense CEL2 glass ceramic (Kohlhauser et al. 2009) (see Table F.2). Samples denoted A-E in this table were cubes with an edge length of about 5 mm, while the rest of the samples collected in Table F.2 were cuboid shaped, with dimensions between 10x10x10 mm and 10x10x50 mm. Equation (F.30) was also used for the estimation of the porosity of the scaffolds soaked in SBF and tris (see Section F.2 for details): this is equivalent to approximating the mass density of the soaking induced, newly formed phases, such as hydroxyapatite with density between 2.61 and 3.16 g/cm^3 in biological systems (Dorozhkin and Epple 2002) by the mass density of CEL2 glass.

F.4.4 Sample specific elasticity experiments on CEL2-based biomaterials – experimental set IIb-1

Elastic properties of porous CEL2-based biomaterials were determined through acoustical testing. The used ultrasonic device is composed of a pulser-receiver Panametrics-NDT 5077 PR, of an oscilloscope, and of several ultrasonic transducers; the pulser unit can emit a square pulse of up to 400 V, with frequencies from 0.1 to 20 MHz. The piezoelectric elements in the transducers are able to transform electrical signals into mechanical ones, or mechanical signals into electrical ones (see Figure F.5).

Speci- men nr.	a measured [mm]	ρ measured [g/cm^3]	φ Eq. (F.30) [%]	v_{bar} Eq. (F.31) [km/s]	λ Eq. (F.32) [mm]	a/λ - -	E_{exp} Eq. (F.33) [GPa]
A	5.22	0.84	67.3	3.96	39.6	0.13	13.10
B	5.35	0.87	66.2	4.09	40.9	0.13	14.50
C	4.33	0.97	62.3	3.94	39.4	0.11	15.00
D	5.22	0.80	68.7	3.06	30.6	0.17	7.50
E	5.14	0.58	77.5	2.97	29.7	0.17	5.10
1	15.27	1.47	42.4	4.71	47.1	0.32	32.73
2	13.34	1.45	43.5	4.31	43.1	0.31	26.87
3	9.78	1.35	47.1	4.09	40.9	0.24	22.61
4	9.74	1.32	48.3	4.16	41.6	0.23	22.85
5	9.85	1.40	45.3	4.08	40.8	0.24	23.28
6	9.59	1.30	49.3	4.08	40.8	0.24	21.58
7	9.5	1.88	26.7	4.73	47.3	0.20	42.02
8	9.5	1.59	37.9	4.43	44.3	0.21	31.13
9	10.39	0.88	65.4	4.34	43.4	0.24	16.70
10	9.74	0.89	65.1	4.24	42.4	0.23	16.10
11	24.75	0.88	65.4	4.25	42.5	0.58	16.00
12	21.6	0.89	65.1	4.13	41.3	0.52	15.30

Table F.2: Porous CEL2-based biomaterial samples: Young's modulus E_{exp} determined from propagation velocity v_{bar} of bar waves with a signal frequency f=0.1 MHz: a is a typical cross-sectional dimension, ρ is the mass density, and φ the porosity of the sample; λ denotes the wavelength

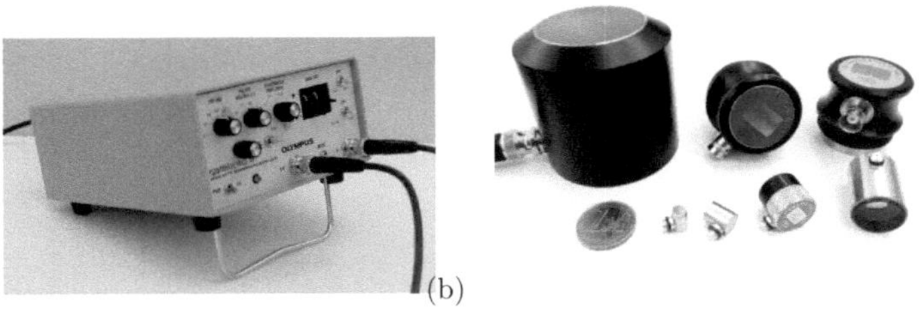

(a) (b)

Figure F.5: Equipment for acoustical testing: (a) pulser-receiver, (b) ultrasonic transducers

The receiver unit has a bandwidth of 0.1-35 MHz and a voltage gain until 59 dB. The signal is displayed on an oscilloscope Lecroy Waverunner 62Xi, which allows for estimating the time of flight t of the acoustic wave through the specimen along a path of length l; t and l give access to the velocity v of the wave, via

$$v = \frac{l}{t} \tag{F.31}$$

Velocity v and frequency f of the acoustic signal yield the wavelength λ as

$$\lambda = \frac{v}{f} \tag{F.32}$$

If the wavelength l is considerably larger than the diameter or another typical cross-sectional dimension a of the specimen, a bar wave propagates with velocity v_{bar} (Fedorov 1968; Ashman et al. 1984). This is the case for the herein employed 0.1 MHz signals propagating through CEL2-based biomaterial samples (see Table F.2). There, the theory of elastodynamics (Fedorov 1968; Ashman et al. 1984) allows for the determination of Young's modulus from the velocities of bar waves

$$E = \rho v_{bar}^2 \tag{F.33}$$

Given $\lambda \approx 40$ mm (see Table F.2) $\gg l_{RVE} = 5$ mm (see Section F.2), these values for Young's modulus actually refer to the (macro)porous biomaterial scaffolds (and not to the dense CEL2 glass ceramic between the macropores).

F.4.5 Comparison between sample specific stiffness predictions and corresponding experiments

The stiffness values predicted by the homogenisation scheme for elastic properties (described in Section F.3) for biomaterial specific porosities (experimental set IIa) on the basis of biomaterial independent ('universal') stiffness of CEL2 biomaterials (experimental set I) are compared to corresponding experimentally determined biomaterial specific stiffness values from experimental set IIb-1. To quantify the model's predictive capabilities, the mean and the standard deviation of the normalised error e, between predictions and experiments $\bar{e}$ and e_S, are considered

$$\bar{e} = \frac{1}{n}\sum_{i=1}^{n} e_i = \frac{1}{n}\sum_{i=1}^{n} e_i \frac{E_{hom,i} - E_{exp}}{\bar{E}_{exp}} \tag{F.34}$$

$$e_S = \left[\frac{1}{n-1}\sum_{i=1}^{n}(e_i - \bar{e})^2\right]^{\frac{1}{2}} \tag{F.35}$$

with summation over n values E_{exp}. $\bar{E}_{exp}$ is the mean over all experimental values.

Insertion of biomaterial specific porosities (fourth column of Table F.2) and 'universal' stiffness constants (Table F.1) into equation (F.18) delivers, together with equation (F.19), sample specific stiffness estimates for the effective Young's modulus E_{hom}. These stiffness predictions are compared to corresponding experimental stiffness values E_{exp} (Figure F.6 and last column of Table F.2). The satisfactory agreement between model predictions and experiments is quantified by prediction errors of $-9 \pm 16\%$ (mean value$\pm$standard deviation).

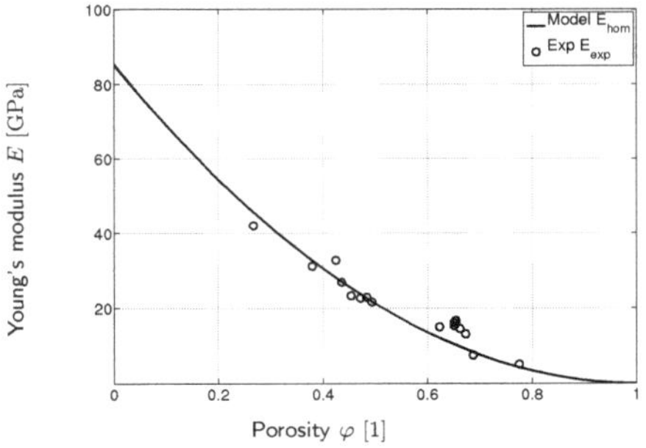

Figure F.6: Comparison between model predictions and experiments for stiffness of porous CEL2 glass ceramic scaffolds

F.4.6 Sample specific strength experiments on CEL2-based biomaterials – experimental set IIb-2

Ultimate properties of CEL2-based biomaterials were determined by uniaxial, compressive, quasistatic testing. The five cubic samples A-E (see also Table F.2 and Section F.4.3) were suitable for measurements in an electromechanical testing stand (MTS QTest 10, see Figure F.7). A 1000 N range force transducer was used. Compression tests were performed in a displacement control mode with 0.015 mm/s speed (strain rate $\sim 3 \cdot 10^{-3}$/s). Corresponding stress strain curves of the specimens are characterised by pronounced softening after a first stress peak. The latter was identified as ultimate strength (see Table F.3).

Figure F.7: Electromechanical testing stand for compression tests on CEL2-based biomaterial samples

Sample	φ [-]	$\Sigma^{ult,c}_{exp}$ [MPa]
A	0.67	1.85
B	0.66	4.58
C	0.62	4.40
D	0.69	2.11
E	0.77	1.91

Table F.3: Experimental compressive strength $\Sigma^{ult,c}_{exp}$ of CEL2-based biomaterial samples as function of porosity φ (see Table F.2)

F.4.7 Comparison between sample specific strength predictions and corresponding experiments

The strength values predicted by the upscaling relations described in Section F.3, for sample specific porosities (experimental set IIa) on the basis of sample independent ('universal') elasticity and shear strength characteristics of dense CEL2 glass ceramic (experimental set I) are compared to corresponding experimentally determined sample specific uniaxial compressive strength values from experimental set IIb-2.

Insertion of biomaterial specific porosities (second column of Table F.3) into equation (F.28), together with equations (F.24), (F.25) and (F.18), delivers, together with E_S, ν_S and τ_S^{ult} (Table F.1), sample specific strength estimates for uniaxial compressive strength ($\Sigma^{ult,c}_{pred}$). These strength predictions are compared to corresponding experimental strength values $\Sigma^{ult,c}_{exp}$ (Figure F.8 and third column of Table F.3). The satisfactory agreement between model predictions and experiments is quantified by prediction errors of 3±34% [mean value±standard deviation, in analogy to equations (F.34)-(F.35)].

F.5 Conclusions

A continuum micromechanical concept has been developed for the elasticity and strength of porous biomaterials made of CEL2, which was verified through independent experimental sets. The latter were gained from the authors' own experiments. The predictions of the porosity based micromechanics model agree well with the corresponding experimentally determined mechanical properties of samples produced by both the replication technique and the burning-out method: this underlines the relevance of the Differential Scheme for microstructures with interconnected pores, irrespective of the actual sphere or strut type microstructural morphology. The good agreement of the model with the corresponding elasticity and strength experiments of samples of both the unmodified and the bioactivity tested biomaterials indicates that the

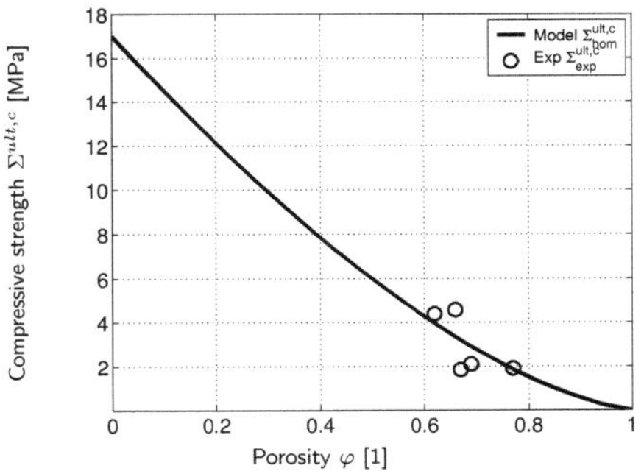

Figure F.8: Comparison between model predictions and experiments for compressive strength of porous CEL2 glass ceramic scaffolds

bioactivity tests primarily increased the porosity of the scaffolds, while the newly formed chemical phases exhibit mechanical properties which are more or less similar to the original glass ceramic phase. The suitability of the differential scheme to predict the elasticity properties of porous CEL2 glass ceramic scaffolds for tissue engineering is consistent with the earlier finding (Zimmermann 1991) that this scheme appropriately predicts the elastic properties of sintered glass (Walsh et al. 1965) of various porosities with nearly spherical pore shape.

Conclusively, it is proposed that micromechanical models have a considerable potential for improving biomaterial design. Nowadays, the latter is largely done in a trial-and-error procedure. Based on a number of mechanical and/or acoustical tests, new material design parameters are guessed. On the other hand, with well validated micromechanical models, the mechanical implications of changes in the microstructure can be predicted so that minimisation of material failure risk allows for the optimisation of key design parameters, such as porosities or geometries of microstructures. Hence, it is believed that micromechanical theories can considerably speed up the future development of tissue engineering scaffolds.

Acknowledgements

This work was supported in part by the EU Network of Excellence project Knowledge-based Multicomponent Materials for Durable and Safe Performance (KMM-NoE) under the contract no. NMP3-CT-2004-502243.

Concluding remarks

We have proposed micromechanical descriptions for different classes of biomaterials, as well as a first multiscale micromechanics model for bone strength, extending earlier developments in the realm of elasticity (Hellmich et al. 2004a; Fritsch and Hellmich 2007).

The employed morphological description for crystals of hydroxyapatite and of bone mineral within biomaterials and bone, respectively, has remarkable analogies to another natural material, namely gypsum. Recent investigations (Sanahuja 2008; Sanahuja et al. 2008) have shown the potential of micromechanical models, based on self-consistent schemes using elongated (prolate) inclusions with different aspect ratios, for predicting elastic and strength properties of gypsum.

As concerns modeling the mechanical properties of bone (Chapter D), such models can potentially support various future scientific as well as application-oriented activities:

1. As was already shown for elasticity (Hellmich et al. 2008), the model is expected to be combined with computer tomographic images: Based on average relations from X-ray physics, the voxel-specific X-ray attenuation information would be translated to voxel-specific material composition; and the latter would serve as input for the micromechanical model, which would then deliver voxel-specific (anisotropic and inhomogeneous) stiffness *and* strength values. In this way, the current activities concerning the virtual physiological human (Taylor et al. 2002; Yosibash et al. 2007; Viceconti et al. 2008), could be extended from the realm of elasticity to that of full elastoplasticity, resulting in patient-specific fracture risk assessment of whole organs in both healthy and pathological conditions.

2. The proposed model could also support the design of tissue engineering scaffolds, through predictions of the failure properties of bone tissue-engineering scaffolds with tissue-engineered bone, by feeding recently developed multiscale representations (Bertrand and Hellmich 2008) not only with an elastic, but with the present elastoplastic micromechanical representation of the extracellular bone material.

3. Since the proposed model is linked to the hierarchical organization of bone and to its elementary components, it is ready to be combined with most recent developments in theoretical and computational biochemistry and biology, which quantify the well-tuned interplay of biological cells via biochemical signaling pathways (Lemaire et al. 2004;

Pivonka et al. 2008) – giving as output the volume fraction of newly deposited or resorbed extravascular bone, which may serve as input for the proposed multiscale strength model. That is expected to open the way to translation of biochemical remodeling events to associated changes in mechanical competence.

As concerns modeling mechanical properties of biomaterials (Chapters B-C and E-F), it is proposed that micromechanical models have a considerable potential for improving biomaterial design. Nowadays, the latter is largely done in a trial-and-error procedure. Based on a number of mechanical and/or acoustical tests, new material design parameters are guessed. On the other hand, with well validated micromechanical models, the mechanical implications of changes in the microstructure can be predicted so that minimisation of material failure risk allows for the optimisation of key design parameters, such as porosities or geometries of microstructures. Hence, it is believed that micromechanical theories can considerably speed up the future development of tissue engineering scaffolds.

Bibliography

Akao, M., Aoki, H., and Kato, K. (1981). Mechanical properties of sintered hydroxyapatite for prosthetic applications. *Journal of Materials Science*, 16:809 – 812.

Akkus, O., Polyakova-Akkus, A., Adar, F., and Schaffler, M. (2003). Aging of microstructural compartments in human compact bone. *Journal of Bone and Mineral Research*, 18(6):1012 – 1019.

Akkus, O. and Rimnac, C. (2001). Cortical bone tissue resists fatigue fracture by deceleration and arrest of microcrack growth. *Journal of Biomechanics*, 34:757 – 764.

Arita, I., Wilkinson, D., Mondragón, M., and Castaño, V. (1995). Chemistry and sintering behaviour of thin hydroxyapatite ceramics with controlled porosity. *Biomaterials*, 16:403 – 408.

Ashman, R., Corin, J., and Turner, C. (1987). Elastic properties of cancellous bone: measurement by an ultrasonic technique. *Journal of Biomechanics*, 20(10):979 – 986.

Ashman, R., Cowin, S., van Buskirk, W., and Rice, J. (1984). A continuous wave technique for the measurement of the elastic properties of cortical bone. *Journal of Biomechanics*, 17(5):349 – 361.

Ballarini, R., Kayacan, R., Ulm, F., Belytschko, T., and Heuer, A. (2005). Biological structures mitigate catastrophic fracture through various strategies. *International Journal of Fracture*, 135:187 – 197.

Baroughel-Bouny, V. (1994). Caractérisation des pâtes de ciment et des bétons - méthodes, analyse, interprétation [Characterization of cement pastes and concretes - methods, analysis, interpretations]. Technical report, Laboratoire Central des Ponts et Chaussées, Paris, France. In French.

Barthelat, F., Tang, H., Zavattieri, P., Li, C.-M., and Espinosa, H. (2007). On the mechanics of mother-of-pearl: A key feature in the material hierarchical structure. *Journal of the Mechanics and Physics of Solids*, 55(2):306 – 337.

Barthélémy, J.-F. and Dormieux, L. (2003). Determination of the macroscopic strength criterion of a porous medium by nonlinear homogenization. *Comptes Rendus Mecanique*, 331:271 – 276.

Barthélémy, J.-F. and Dormieux, L. (2004). A micromechanical approach to the strength criterion of Drucker-Prager materials reinforced by rigid inclusions. *International Journal for Numerical and Analytical Methods in Geomechanics*, 28(7-8):565 – 582.

Benezra Rosen, V., Hobbs, L., and Spector, M. (2002). The ultrastructure of anorganic bovine bone and selected synthetic hydroxyapatites used as bone graft substitute material. *Biomaterials*, 23:921 – 928.

Benveniste, Y. (1987). A new approach to the application of Mori-Tanaka's theory in composite materials. *Mechanics of Materials*, 6:147 – 157.

Bernard, O., Ulm, F., and Lemarchand, E. (2003). A multiscale micromechanics – hydration model for the early-age elastic properties of cement-based materials. *Cement and Concrete Research*, 33(9):1293 – 1309.

Bernaud, D., Deudé, V., Dormieux, L., Maghous, S., and Schmitt, D. (2002). Evolution of properties in finite poroplasticity and finite element analysis. *International Journal for Numerical and Analytical Methods in Geomechanics*, 26:845 – 871.

Bertrand, E. and Hellmich, C. (2008). Multiscale elasticity of tissue engineering scaffolds with tissue-engineered bone: a continuum micromechanics approach. *Journal of Engineering Mechanics*. In print.

Bhowmik, R., Katti, K., and Katti, D. (2007). Mechanics of molecular collagen is influenced by hydroxyapatite in natural bone. *Journal of Materials Science*, 42:8795 – 8803.

Boccaccini, A., Blazer, J., Maquet, V., Day, R., and Jérôme, R. (2005). Preparation and characterisation of poly(lactide-co-glycolide) (PLGA) and PLGA/Bioglass® composite tubular foam scaffolds for tissue engineering applications. *Materials Science and Engineering C*, 25:23 – 31.

Boivin, G. and Meunier, P. (2002). The degree of mineralization of bone tissue measured by computerized quantitative contact microradiography. *Calcified Tissue International*, 70:503 – 511.

Bonar, L., Lees, S., and Mook, H. (1985). Neutron diffraction studies of collagen in fully mineralized bone. *Journal of Molecular Biology*, 181:265 – 270.

Boskey, A. (2003). Bone mineral crystal size. *Osteoporosis International*, 14(Suppl 5):S16 – S21.

Bossy, E., Talmant, M., Peyrin, F., Akrout, L., Cloetens, P., and Laugier, P. (2004). In in vitro study of the ultrasonic axial transmission technique at the radius: 1 MHz velocity measurements are sensitive to both mineralization and introcortical porosity. *Journal of Bone and Mineral Research*, 19(9):1548 – 1556.

Boucher, S. (1976). Modules effectifs de matériaux composites quasi homogènes et quasi isotropes, constitués d'une matrice élastique et d'inclusions élastiques. II. Cas des concentrations finies en inclusions. *Revue M*, 22(1):31 – 36.

Bousson, V., Bergot, C., Meunier, A., Barbot, F., Parlier-Cuau, C., Laval-Jeantet, A.-M., and Laredo, J.-D. (2000). CT of the middiaphyseal femur: Cortical bone mineral density and relation to porosity. *Radiology*, 217:179–187.

Buehler, M. (2006). Nature designs tough collagen: Explaining the nanostructure of collagen fibrils. *Proceedings of the National Academy of Sciences of the United States of America (PNAS)*, 103(33):12285 – 12290.

Buehler, M. (2008). Nanomechanics of collagen fibrils under varying cross-link densities: Atomistic and continuum studies. *Journal of the Mechanical Behavior of Biomedical Materials*, 1:59 – 67.

Burr, D., Turner, C., Naick, P., Forwood, M., Ambrosius, W., Hasan, M., and Pidaparti, R. (1998). Does microdamage accumulation affect the mechanical properties of bone. *Journal of Biomechanics*, 31:337 – 345.

Burstein, A., Currey, J., Frankel, V., and Reilly, D. (1972). The ultimate properties of bone tissue: The effects of yielding. *Journal of Biomechanics*, 5:35 – 44.

Burstein, A., Reilly, D., and Martens, M. (1976). Aging of bone tissue: Mechanical properties. *Journal of Bone and Joint Surgery*, 58A:82 – 86.

Burstein, A., Zika, J., Heiple, K., and Klein, L. (1975). Contribution of collagen and mineral to the elastic-plastic properties of bone. *Journal of Bone and Joint Surgery*, 57A:956 – 961.

Buttery, L. and Bishop, A. (2005). Introduction to tissue engineering. In Hench, L. and Jones, J., editors, *Biomaterials, artificial organs and tissue engineering*, pages 193 – 200. Woodhead Publishing, Cambridge, U.K.

Cancedda, R., Cedola, A., Giuliani, A., Komlev, V., Lagomarsino, S., Mastrogiacomo, M., Peyrin, F., and Rustichelli, F. (2007). Bulk and interface investigations of scaffolds and tissue-engineered bones by X-ray microtomography and X-ray microdiffraction. *Biomaterials*, 28:2505 – 2524.

Carcione, J. (2001). *Wave fields in real media: Wave propagation in anisotropic, anelastic and porous media*. Pergamon, Oxford, UK.

Catanese, J., Iverson, E., Ng, R., and Keaveny, T. (1999). Heterogeneity of the mechanical properties of demineralized bone. *Journal of Biomechanics*, 32:1365 – 1369.

Cezayirlioglu, H., Bahniuk, E., Davy, D., and Heiple, K. (1985). Anisotropic yield behavior of bone under combined axial force and tension. *Journal of Biomechanics*, 18(1):61 – 69.

Charrière, E., Terrazzoni, S., Pittet, C., Mordasini, P., Dutoit, M., Lemaître, J., and Zysset, P. (2001). Mechanical characterization of brushite and hydroxyapatite cements. *Biomaterials*, 22:2937 – 2945.

Chateau, X. and Dormieux, L. (2002). Micromechanics of saturated and unsaturated porous media. *International Journal for Numerical and Analytical Methods in Geomechanics*, 26:831 – 844.

Christiansen, D., Huang, E., and Silver, F. (2000). Assembly of type I collagen: fusion of fibril subunits and the influence of fibril diameter on mechanical properties. *Matrix Biology*, 19:409 – 420.

Chu, T.-M., Orton, D., Hollister, S., Feinberg, S., and Halloran, J. (2002). Mechanical and in vivo performance of hydroxyapatite implants with controlled architecture. *Biomaterials*, 23:1283 – 1293.

Coussy, O. (2004). *Poromechanics*. Wiley, Chichester, NJ.

Cowin, S. (2003). A recasting of anisotropic poroelasticity in matrices of tensor components. *Transport in Porous Media*, 50:35 – 56.

Cowin, S. and Mehrabadi, M. (1992). The structure of the linear anisotropic elastic symmetries. *Journal of the Mechanics and Physics of Solids*, 40:1459 – 1471.

Crank, J. (1975). *The mathematics of diffusion*. Oxford Science Publications, New York, USA, 2 edition.

Currey, J. (1959). Differences in the tensile strength of bone of different histological types. *Journal of Anatomy*, 93:87 – 95.

Currey, J. (1969). The relationship between the stiffness and the mineral content of bone. *Journal of Biomechanics*, 2:477 – 480.

Currey, J. (1975). The effects of strain rate, reconstruction and mineral content on some mechanical properties of bovine bone. *Journal of Biomechanics*, 8:81 – 86.

Currey, J. (1984). Effects of differences in mineralization on the mechanical properties of bone. *Philosophical Transactions of the Royal Society of London, Series B*, 304:509 – 518.

Currey, J. (1988). Strain rate and mineral content in fracture models of bone. *Journal of Orthopaedic Research*, 6(1):32 – 38.

Currey, J. (1990). Physical characteristics affecting the tensile failure properties of compact bone. *Journal of Biomechanics*, 23:837 – 844.

Currey, J. (2004). Tensile yield in compact bone is determined by strain, post-yield behaviour by mineral content. *Journal of Biomechanics*, 37:549 – 556.

Cusack, S. and Miller, A. (1979). Determination of the elastic constants of collagen by Brillouin light scattering. *Journal of Molecular Biology*, 135:39 – 51.

De With, G., van Dijk, H., Hattu, N., and Prijs, K. (1981). Preparation, microstructure and mechanical properties of dense polycrystalline hydroxy apatite. *Journal of Materials Science*, 16:1592 – 1598.

Dickenson, R., Hutton, W., and Stott, J. (1981). The mechanical properties of bone in osteoporosis. *Journal of Bone and Joint Surgery*, 63-B(2):233 – 238.

Ding, M. and Hvid, I. (2000). Quantification of age-related changes in the structure model type and trabecular thickness of human tibial cancellous bone. *Bone*, 26:291 – 295.

Dormieux, L. (2005). Poroelasticity and strength of fully or partially saturated porous materials. In Dormieux, L. and Ulm, F.-J., editors, *CISM Vol. 480 – Applied Micromechanics of Porous Media*, pages 109 – 152. Springer, Wien.

Dormieux, L., Barthélémy, J.-F., and Maghous, S. (2006a). Résistance d'un composite à renforts rigides: le cas d'une matrice de Drucker-Prager avec règle d'écoulement plastique non associée [Strength of a composite reinforced by rigid inclusions: the case of a Drucker-Prager matrix with non associated plastic flow rule]. *Comptes Rendus Mecanique*, 334:111 – 116. In French.

Dormieux, L., Kondo, D., and Ulm, F.-J. (2006b). *Microporomechanics*. Wiley.

Dormieux, L. and Lemarchand, E. (2001). Homogenization approach of advection and diffusion in cracked porous material. *Journal of Engineering Mechanics (ASCE)*, 127(8):1267 – 1275.

Dormieux, L. and Maghous, S. (2000). Evolution des propriétés élastiques en poroplasticité finie [Evolution of elastic properties in finite poroplasticity]. *Comptes Rendus de l'Académie des Sciences – Series IIb – Mechanics*, 328(8):593 – 600. In French.

Dormieux, L., Molinari, A., and Kondo, D. (2002). Micromechanical approach to the behavior of poroelastic materials. *Journal of the Mechanics and Physics of Solids*, 50:2203 – 2231.

Dormieux, L., Sanahuja, J., and Maalej, Y. (2007). Résistance d'un polycrystal avec interfaces intergranulaires imparfaites [Strength of a polycrystal with imperfect intergranular interfaces]. *Comptes Rendus Mecanique*, 335(1):25 – 31.

Dormieux, L., Sanahuja, J., and Maghous, S. (2006c). Influence of capillary effects on strength of non-saturated porous media. *Comptes Rendus Mecanique*, 334:19 – 24.

Dorozhkin, S. and Epple, M. (2002). Biological and medical significance of calcium phosphates. *Angewandte Chemie International Edition*, 41:3130 – 3146.

Driessen, A., Klein, C., and de Groot, K. (1982). Preparation and some properties of sintered β-whitlockite. *Biomaterials*, 3:113 – 116.

Dvorak, G. (1992). Transformation field analysis of inelastic composite materials. *Proceedings of the Royal Society London, Series A*, 437:311 – 327.

Eppell, S., Tong, W., Katz, J., Kuhn, L., and Glimcher, M. (2001). Shape and size of isolated bone mineralites measured using atomic force microscopy. *Journal of Orthopaedic Research*, 19:1027 – 1034.

Eshelby, J. (1957). The determination of the elastic field of an ellipsoidal inclusion, and related problems. *Proceedings of the Royal Society London, Series A*, 241:376 – 396.

Fedorov, F. (1968). *Theory of elastic waves in crystals*. Plenum Press, New York.

Frame, J., Browne, R., and Brady, C. (1981). Hydroxyapatite as a bone substitute in the jaws. *Biomaterials*, 2:19 – 22.

Fratzl, P., Fratzl-Zelman, N., Klaushofer, K., Vogl, G., and Koller, K. (1991). Nucleation and growth of mineral crystals in bone studied by small-angle X-ray scattering. *Calcified Tissue International*, 48:407 – 413.

Fratzl, P., Schreiber, S., and Klaushofer, K. (1996). Bone mineralization as studied by small-angle X-ray scattering. *Connective Tissue Research*, 34(4):247 – 254.

Fritsch, A., Dormieux, L., and Hellmich, C. (2006). Porous polycrystals built up by uniformly and axisymmetrically oriented needles: Homogenization of elastic properties. *Comptes Rendus Mécanique*, 334(3):151 – 157.

Fritsch, A., Dormieux, L., Hellmich, C., and Sanahuja, J. (2007a). Micromechanics of crystal interfaces in polycrystalline solid phases of porous media: fundamentals and application to strength of hydroxyapatite biomaterials. *Journal of Materials Science*, 42(21):8824 – 8837.

Fritsch, A., Dormieux, L., Hellmich, C., and Sanahuja, J. (2007b). Micromechanics of hydroxyapatite-based biomaterials and tissue engineering scaffolds. In Boccaccini, A. and

Gough, J., editors, *Tissue engineering using ceramics and polymers*, chapter 25, pages 529 – 565. Woodhead Publishing, Cambridge, U.K.

Fritsch, A., Dormieux, L., Hellmich, C., and Sanahuja, J. (2009a). Mechanical behaviour of hydroxyapatite biomaterials: An experimentally validated micromechanical model for elasticity and strength. *Journal of Biomedical Materials Research Part A*, 88A:149 – 161.

Fritsch, A. and Hellmich, C. (2007). 'Universal' microstructural patterns in cortical and trabecular, extracellular and extravascular bone materials: Micromechanics-based prediction of anisotropic elasticity. *Journal of Theoretical Biology*, 244(4):597 – 620.

Fritsch, A., Hellmich, C., and Dormieux, L. (2009b). Ductile sliding between mineral crystals followed by rupture of collagen crosslinks: experimentally supported micromechanical explanation of bone strength. *Journal of Theoretical Biology*. Submitted for publication.

Fung, Y. (2002). Celebrating the inauguration of the journal: Biomechanics and Modeling in Mechanobiology. *Biomechanics and Modeling in Mechanobiology*, 1:3 – 4.

Gennes, P.-G. d. and Okumura, K. (2000). On the toughness of biocomposites. *Comptes Rendus de l'Académie des Sciences-Series IV*, 1(2):257 – 261.

Gentleman, E., Lay, A., Dickerson, D., Nauman, E., Livesay, G., and Dee, K. (2003). Mechanical characterization of collagen fibers and scaffolds for tissue engineering. *Biomaterials*, 24:3805 – 3813.

Gibson, L. (1985). The mechanical behavior of cancellous bone. *Journal of Biomechanics*, 18:317 – 28.

Gibson, L. and Ashby, M. (1997). *Cellular Solids: Structure and Properties*. Cambridge University Press, Cambridge, UK, 2 edition.

Gilmore, R. and Katz, J. (1982). Elastic properties of apatites. *Journal of Materials Science*, 17:1131 – 1141.

Gould, S. and Lewontin, R. (1979). The spandrels of San Marco and the Panglossian paradigm: a critique of the adaptionist program. *Proceedings of the Royal Society of London, Series B*, 205(1161):581 – 598.

Grimm, M. and Williams, J. (1997). Measurements of permeability in human calcaneal trabecular bone. *Journal of Biomechanics*, 30:743 – 745.

Gruescu, C., Monchiet, V., and Kondo, D. (2005). Eshelby tensor for a crack in an orthotropic elastic medium. *Comptes Rendus Mecanique*, 333(6):467 – 473.

Hashin, Z. (1983). Analysis of composite materials: a survey. *Journal of Applied Mechanics*, 50:481 – 505.

Hashin, Z. (1991). A spherical inclusion with imperfect interface. *Journal of Applied Mechanics*, 58:444 – 449.

Hassenkam, T., Fantner, G., Cutroni, J., Weaver, J., Morse, D., and Hansma, P. (2004). High-resolution AFM imaging of intact and fractured trabecular bone. *Bone*, 35:4 – 10.

Hellmich, C. (2004). Microelasticity of bone. In Dormieux, L. and Ulm, F.-J., editors, *CISM Courses and Lectures, vol. 480. Applied Micromechanics of Porous Media*, pages 289 – 332. Springer, Wien – New York.

Hellmich, C., Barthélémy, J.-F., and Dormieux, L. (2004a). Mineral-collagen interactions in elasticity of bone ultrastructure – a continuum micromechanics approach. *European Journal of Mechanics A-Solids*, 23:783 – 810.

Hellmich, C., Kober, C., and Erdmann, B. (2008). Micromechanics-based conversion of CT data into anisotropic elasticity tensors, applied to FE simulations of a mandible. *Annals of Biomedical Engineering*, 36:108 – 122.

Hellmich, C. and Mang, H. (2005). Shotcrete elasticity revisited in the framework of continuum micromechanics: From submicron to meter level. *Journal of Materials in Civil Engineering (ASCE)*, 17(3):246–256.

Hellmich, C., Müllner, H., and Kohlhauser, C. (2006). Mechanical (triaxial) tests on biological materials and biomaterials. Technical Report DNRT3-1.2-3, Network of Excellence 'Knowledge-based Multicomponent Materials for Durable and Safe Performance – KMM-NoE', sponsored by the European Commission.

Hellmich, C. and Ulm, F.-J. (2001). Hydroxyapatite is uniformly concentrated in the extracollagenous ultrastructure of mineralized tissue. In Middleton, J., Shrive, N., and Jones, M., editors, *Proceedings of the Fifth International Symposium on Computer Methods in Biomechanics and Biomedical Engineering*, Rome, Italy.

Hellmich, C. and Ulm, F.-J. (2002a). Are mineralized tissues open crystal foams reinforced by crosslinked collagen? – some energy arguments. *Journal of Biomechanics*, 35:1199 – 1212.

Hellmich, C. and Ulm, F.-J. (2002b). A micromechanical model for the ultrastructural stiffness of mineralized tissues. *Journal of Engineering Mechanics (ASCE)*, 128(8):898 – 908.

Hellmich, C. and Ulm, F.-J. (2003). Average hydroxyapatite concentration is uniform in extracollageneous ultrastructure of mineralized tissue. *Biomechanics and Modeling in Mechanobiology*, 2:21 – 36.

Hellmich, C. and Ulm, F.-J. (2005a). Drained and undrained poroelastic properties of healthy and pathological bone: a poro-micromechanical investigation. *Transport in Porous Media*, 58:243 – 268.

Hellmich, C. and Ulm, F.-J. (2005b). Micro-porodynamics of bones: prediction of the 'Frenkel-Biot' slow compressional wave. *Journal of Engineering Mechanics (ASCE)*, 131(9):918 – 927.

Hellmich, C., Ulm, F.-J., and Dormieux, L. (2004b). Can the diverse elastic properties of trabecular and cortical bone be attributed to only a few tissue-independent phase properties and their interactions? – Arguments from a multiscale approach. *Biomechanics and Modeling in Mechanobiology*, 2:219 – 238.

Hench, L. and Jones, J., editors (2005). *Biomaterials, artificial organs and tissue engineering.* Woodhead Publishing, Cambridge, U.K.

Hernandez, C., Beaupré, G., Keller, T., and Carter, D. (2001). The influence of bone volume fraction and ash fraction on bone strength and modulus. *Bone*, 29:74 – 78.

Hershey, A. (1954). The elasticity of an isotropic aggregate of anisotropic cubic crystals. *Journal of Applied Mechanics (ASME)*, 21:236 – 240.

Hervé, E. and Zaoui, A. (1993). n-layered inclusion-based micromechanical modelling. *International Journal of Engineering Science*, 31(1):1 – 10.

Hill, R. (1963). Elastic properties of reinforced solids: some theoretical principles. *Journal of the Mechanics and Physics of Solids*, 11:357 – 362.

Hill, R. (1965). Continuum micro-mechanics of elastoplastic polycrystals. *Journal of the Mechanics and Physics of Solids*, 13:89 – 101.

Hofstetter, K., Hellmich, C., and Eberhardsteiner, J. (2005). Development and experimental validation of a continuum micromechanics model for the elasticity of wood. *European Journal of Mechanics A - Solids*, 24(6):1030 – 1053.

Hofstetter, K., Hellmich, C., and Eberhardsteiner, J. (2006). The influence of the microfibril angle on wood stiffness: a continuum micromechanics approach. *Computer Assisted Mechanics and Engineering Sciences*, 13(4):523 – 536.

Hosoda, H., Kinoshita, Y., Fukui, Y., Inamura, T., Wakashima, K., Kim, H., and Miyazaki, S. (2006). Effects of short time heat treatment on superelastic properties of a TiNbAl biomedical shape memory alloy. *Materials Science and Engineering A*, 438-440:870 – 874.

Hunt, B., Eipsman, R., and Rosenberg, J. (2001). *A Guide to MATLAB for Beginners and Experienced Users.* Cambridge University Press, Cambridge, United Kingdom, 1 edition.

Hunter, G., Hauschka, P., Poole, A., Rosenberg, L., and Goldberg, H. (1996). Nucleation and inhibition of hydroxyapatite formation by mineralized tissue proteins. *Biochemical Journal*, 317:59 – 64.

Jäger, I. and Fratzl, P. (2000). Mineralized collagen fibrils: a mechanical model with a staggered arrangement of mineral particles. *Biophysical Journal*, 79:1737 – 1746.

Jones, J. (2005). Scaffolds for tissue engineering. In Hench, L. and Jones, J., editors, *Biomaterials, artificial organs and tissue engineering*, pages 201 – 214. Woodhead Publishing, Cambridge, U.K.

Katti, D. and Katti, K. (2001). Modeling microarchitecture and mechanical behavior of nacre using 3D finite element techniques. Part I. Elastic properties. *Journal of Materials Science*, 36(6):1411 – 1417.

Katti, D., Katti, K., Sopp, J., and Sarikaya, M. (2001). 3D finite element modeling of mechanical response in nacre-based hybrid nanocomposites. *Computational and Theoretical Polymer Science*, 11(5):397 – 404.

Katz, E. and Li, S.-T. (1973). Structure and function of bone collagen fibrils. *Journal of Molecular Biology*, 80:1 – 15.

Katz, J. (1980). Anisotropy of Young's modulus of bone. *Nature*, 283:106 – 107.

Katz, J. (1981). *Composite material models for cortical bone*, pages 171 – 184. American Society of Mechanical Engineers, New York, NY, USA.

Katz, J. and Harper, R. (1990). Calcium phosphates and apatites. In Williams, D., editor, *Concise encyclopedia of medical and dental materials*, pages 87 – 95. Pergamon Press, Oxford, U.K.

Katz, J. and Ukraincik, K. (1971). On the anisotropic elastic properties of hydroxyapatite. *Journal of Biomechanics*, 4:221 – 227.

Katz, J., Yoon, H., Lipson, S., Maharidge, R., Meunier, A., and Christel, P. (1984). The effects of remodelling on the elastic properties of bone. *Calcified Tissue International*, 36:S31 – S36.

Keaveny, T., Borchers, R., Gibson, L., and Hayes, W. (1993). Trabecular bone modulus and strength can depend on specimen geometry. *Journal of Biomechanics*, 26(8):991 – 1000.

Kobayashi, K., Sakai, J., and Sakamoto, M. (2001). Strength criterion for bovine tibial trabecular bone under multiaxial stress. *Theoretical and Applied Mechanics*, 50:91 – 96.

Koester, K., Ager, J., and Ritchie, R. (2008). The true toughness of human cortical bone measured with realistically short cracks. *Nature Materials*, 7:672 – 677.

Kohlhauser, C., Hellmich, C., Vitale-Brovarone, C., Boccaccini, A., Rota, A., and Eberhardsteiner, J. (2009). Ultrasonic characterisation of porous biomaterials across different frequencies. *Strain*, 45:34 – 44.

Koiter, W. (1960). General theorems for elastic-plastic solids. In Sneddon, I. and Hill, R., editors, *Progress in solid mechanics*, volume I, chapter IV, pages 167 – 218. North-Holland Publishing Company, Amsterdam, The Netherlands.

Kolsky, H. (1953). *Stress waves in solids.* Clarendon Press, Oxford, UK, Cambridge, United Kingdom.

Kotha, S. and Guzelsu, N. (2002). Modeling the tensile mechanical behavior of bone along the longitudinal direction. *Journal of Theoretical Biology*, 219:269 – 279.

Kotha, S. and Guzelsu, N. (2003). Effect of bone mineral content on the tensile properties of cortical bone: Experiments and theory. *Journal of Biomechanical Engineering*, 125:785 – 793.

Kreher, W. (1990). Residual stresses and stored elastic energy of composites and polycrystals. *Journal of the Mechanics and Physics of Solids*, 38(1):115 – 128.

Kreher, W. and Molinari, A. (1993). Residual stresses in polycrystals as influenced by grain shape and texture. *Journal of the Mechanics and Physics of Solids*, 41(12):1955 – 1977.

Kröner, E. (1958). Computation of the elastic constants of polycrystals from constants of single crystals. *Zeitschrift für Physik*, 151:504 – 518. In German.

Langer, R. and Vacanti, J. (1993). Tissue engineering. *Science*, 260:920 – 926.

Laws, N. (1977). The determination of stress and strain concentrations at an ellipsoidal inclusion in an anisotropic material. *Journal of Elasticity*, 7(1):91 – 97.

Laws, N. (1985). A note on penny-shaped cracks in transversely isotropic materials. *Mechanics of Materials*, 4:209 – 212.

Lee, S., Coan, B., and Bouxsein, M. (1997). Tibial ultrasound velocity measured in situ predicts the material properties of tibial cortical bone. *Bone*, 21(1):119 – 125.

Lees, S. (1987a). Considerations regarding the structure of the mammalian mineralized osteoid from viewpoint of the generalized packing model. *Connective Tissue Research*, 16:281 – 303.

Lees, S. (1987b). Possible effect between the molecular packing of collagen and the composition of bony tissues. *International Journal of Biological Macromolecules*, 9:321 – 326.

Lees, S., Ahern, J., and Leonard, M. (1983). Parameters influencing the sonic velocity in compact calcified tissues of various species. *Journal of the Acoustical Society of America*, 74(1):28 – 33.

Lees, S., Bonar, L., and Mook, H. (1984a). A study of dense mineralized tissue by neutron diffraction. *International Journal of Biological Macromolecules*, 6:321 – 326.

Lees, S., Cleary, P., Heeley, J., and Gariepy, E. (1979a). Distribution of sonic plesio-velocity in a compact bone sample. *Journal of the Acoustical Society of America*, 66(3):641– 646.

Lees, S., Hanson, D., and Page, E. (1995). Some acoustical properties of the otic bones of a fin whale. *Journal of the Acoustical Society of America*, 99(4):2421 – 2427.

Lees, S., Heeley, J., and Cleary, P. (1979b). A study of some properties of a sample of bovine cortical bone using ultrasound. *Calcified Tissue International*, 29:107 – 117.

Lees, S. and Page, E. (1992). A study of some properties of mineralized turkey leg tendon. *Connective Tissue Research*, 28:263 – 287.

Lees, S., Pineri, M., and Escoubes, M. (1984b). A generalized packing model for type I collagen. *International Journal of Biological Macromolecules*, 6:133 – 136.

Lees, S., Prostak, K., Ingle, V., and Kjoller, K. (1994). The loci of mineral in turkey leg tendon as seen by atomic force microscope and electron microscopy. *Calcified Tissue International*, 55:180 – 189.

Lees, S., Tao, N.-J., and Lindsay, M. (1990). Studies of compact hard tissues and collagen by means of Brillouin light scattering. *Connective Tissue Research*, 24:187 – 205.

LeGeros, R. (2002). Properties of osteoconductive biomaterials: Calcium phosphates. *Clinical Orthopaedics and Related Research*, 395:81 – 98.

Lemaire, V., Tobin, F., Greller, L., Cho, C., and Suva, L. (2004). Modeling the interactions between osteoblast and osteoclast activities in bone remodeling. *Journal of Theoretical Biology*, 229:293 – 309.

Lemarchand, E., Ulm, F.-J., and Dormieux, L. (2002). Effect of inclusions on friction coefficient of highly filled composite materials. *Journal of Engineering Mechanics (ASCE)*, 128(8):876 – 884.

Leong, K. and Jin, L. (2006). Characteristics of oscillating flow through a channel filled with open-cell metal foam. *International Journal of Heat and Fluid Flow*, 27:144 – 153.

Levin, V., Michelitsch, T., and Sevostianov, I. (2000). Spheroidal inhomogeneity in a transversely isotropic piezoelectric medium. *Archive of Applied Mechanics*, 70:673 – 693.

Li, G., Bronk, J., An, K.-N., and Kelly, P. (1987). Permeability of cortical bone of canine tibiae. *Microvascular Research*, 34:302 – 310.

Lim, T.-H. and Hong, J. (2000). Poroelastic properties of bovine vertebral trabecular bone. *Journal of Orthopaedic Research*, 18:671 – 677.

Liu, D.-M. (1997). Fabrication of hydroxyapatite ceramic with controlled porosity. *Journal of Materials Science: Materials in Medicine*, 8:227 – 232.

Liu, D.-M. (1998). Preparation and characterisation of porous hydroxyapatite bioceramic via a slip-casting route. *Ceramics International*, 24:441 – 446.

Malasoma, A., Fritsch, A., Kohlhauser, C., Brynk, T., Vitale-Brovarone, C., Pakiela, Z., Eberhardsteiner, J., and Hellmich, C. (2008). Micromechanics of bioresorbable porous CEL2 glass-ceramic scaffolds for bone tissue engineering. *Advances in Applied Ceramics*, 107:277 – 286.

Mammone, J. and Hudson, S. (1993). Micromechanics of bone strength and failure. *Journal of Biomechanics*, 26:439 – 446.

Martin, R. and Brown, P. (1995). Mechanical properties of hydroxyapatite formed at physiological temperature. *Journal of Materials Science: Materials in Medicine*, 6:138 – 143.

Martin, R., Burr, D., and Sharkey, N. (1998). *Skeletal Tissue Mechanics*. Springer, New York.

Martin, R. and Ishida, J. (1989). The relative effects of collagen fiber orientation, porosity, density, and mineralization on bone strength. *Journal of Biomechanics*, 22:419 – 426.

Mastrogiacomo, M., Scaglione, S., Martinetti, R., Dolcini, L., Beltrame, F., Cancedda, R., and Quarto, R. (2006). Role of scaffold internal structure on in vivo bone formation in macroporous calcium phosphate bioceramics. *Biomaterials*, 27:3230 – 3237.

Matweb (2007). Automations Creations Inc., Material type: Titanium alloy.

Mayr, E. (1997). *This is biology – the science of the living world*. Harvard University Press, Cambridge, MA, USA, 1st edition.

McCalden, R., McGeough, J., Barker, M., and Court-Brown, C. (1993). Age-related changes in the tensile properties of cortical bone. The relative importance of changes in porosity, mineralization, and microstructure. *Journal of Bone and Joint Surgery*, 75-A(8):1193 – 1205.

McLaughlin, R. (1977). A study of the differential scheme for composite materials. *International Journal of Engineering Science*, 15:237 – 244.

McNeil, D. and Stuart, A. (2004). Vertically upward two-phase flow with a highly viscous liquid-phase in a nozzle and orifice plate. *International Journal of Heat and Fluid Flow*, 25:58 – 73.

Miller, A. (1984). Collagen: the organic matrix of bone. *Philosophical Transactions of the Royal Society in London Series B*, 304:455 – 477.

Mizuno, S., Watanabe, S., and Takagi, T. (2004). Hydrostatic fluid pressure promotes cellularity and proliferation of human dermal fibroblasts in a three-dimensional collagen gel/sponge. *Biochemical Engineering Journal*, 20:203 – 208.

Molinari, A. and El Mouden, M. (1996). The problem of elastic inclusions at finite concentration. *International Journal of Solids and Structures*, 33:3131 – 3150.

Morgan, E., Lee, J., and Keaveny, T. (2005). Sensitivity of multiple damage parameters to compressive overload in cortical bone. *Journal of Biomedical Engineering*, 127:557 – 562.

Mori, T. and Tanaka, K. (1973). Average stress in matrix and average elastic energy of materials with misfitting inclusions. *Acta Metallurgica*, 21(5):571 – 574.

Müllner, H., Fritsch, A., Kohlhauser, C., Reihsner, R., Hellmich, C., Godlinski, D., Rota, A., Slesinski, R., and Eberhardsteiner, J. (2008). Acoustical and poromechanical characterization of titanium scaffolds for biomedical applications. *Strain*, 44:153 – 163.

Nalla, R., Kruzic, J., and Ritchie, R. (2004). On the origin of the toughness of mineralized tissue: microcracking or crack bridging? *Bone*, 34:790 – 798.

Nowlan, N. and Prendergast, P. (2005). Evolution of mechanoregulation of bone growth will lead to non-optimal bone phenotypes. *Journal of Theoretical Biology*, 235:408 – 418.

Nyman, J., Ni, Q., Nicolella, D., and Wang, X. (2008). Measurements of mobile and bound water by nuclear magnetic resonance correlate with mechanical properties of bone. *Bone*, 42:193 – 199.

O'Brien, F., Taylor, D., and Lee, T. (2007). Bone as a composite material: The role of osteons as barriers to crack growth in compact bone. *International Journal of Fatigue*, 29:1051 – 1056.

Ochoa, J., Sanders, A., Heck, D., and Hilberry, B. (1991). Stiffening of the femoral head due to intertrabecular fluid and intraosseous pressure. *Journal of Biomechanical Engineering*, 113:259 – 261.

Okumura, K. (2002). Why is nacre strong? II. Remaining mechanical weakness for cracks propagating along the sheets. *The European Physical Journal E - Soft Matter*, 7(4):303 – 310.

Okumura, K. (2003). Enhanced energy of parallel fractures in nacre-like composite materials. *Europhysics Letters*, 63(5):701 – 707.

Okumura, K. and Gennes, P.-G. d. (2001). Why is nacre strong? Elastic theory and fracture mechanics for biocomposites with stratified structures. *The European Physical Journal E - Soft Matter*, 4(1):121 – 127.

Pan, H., Tao, J., Wu, T., and Tang, R. (2007). Molecular simulation of water behaviors on crystal faces of hydroxyapatite. *Frontiers of Chemistry in China*, 2:156 – 163.

Peelen, J., Rejda, B., and de Groot, K. (1978). Preparation and properties of sintered hydroxylapatite. *Ceramurgia International*, 4(2):71 – 74.

Peters, F., Schwarz, K., and Epple, M. (2000). The structure of bone studied with synchrotron X-ray diffraction, X-ray absorption spectroscopy and thermal analysis. *Thermochimica Acta*, 361:131 – 138.

Pichler, B., Hellmich, C., and Dormieux, L. (2007a). Potentials and limitations of Griffith's energy release rate criterion for mode I type microcracking in brittle materials. In Exadaktylos, G. and Vardoulakis, I., editors, *Bifurcations, Instabilities, Degradation in Geomechanics*, pages 245 – 276. Springer, Berlin.

Pichler, B., Hellmich, C., and Eberhardsteiner, J. (2008a). Spherical and acicular representation of hydrates in a micromechanical model for cement paste: prediction of early-age elasticity and strength. *Acta Mechanica*. In print, available online at www.springerlink.com, doi:10.1007/s00707-008-0007-9.

Pichler, B., Hellmich, C., and Mang, H. (2007b). A combined fracture-micromechanics model for tensile strain-softening in brittle materials, based on propagation of interacting microcracks. *International Journal for Numerical and Analytical Methods in Geomechanics*, 31:111 – 132.

Pichler, B., Scheiner, S., and Hellmich, C. (2008b). From micron-sized needle-shaped hydrates to meter-sized shotcrete tunnel shells: Micromechanical upscaling of stiffness and strength of shotcrete. *Acta Geotechnica*, 3:273 – 294.

Pidaparti, R., Chandran, A., Takano, Y., and Turner, C. (1996). Bone mineral lies mainly outside the collagen fibrils: Predictions of a composite model for osteonal bone. *Journal of Biomechanics*, 29(7):909 – 916.

Pidaparti, R., Merril, B., and Downton, N. (1997). Fracture and material degradation properties of cortical bone under accelerated stress. *Journal of Biomedical Materials Research*, 37:161 – 165.

Pivonka, P., Zimak, J., Smith, D., Gardiner, B., Dunstan, C., Sims, N., Martin, T., and Mundy, G. (2008). Model structure and control of bone remodeling: A theoretical study. *Bone*, 43:249 – 263.

Pramanik, S., Agarwal, A., Rai, K., and Garg, A. (2007). Development of high strength hydroxyapatite by solid-state-sintering process. *Ceramics International*, 33(3):419 – 426.

Prostak, K. and Lees, S. (1996). Visualization of crystal-matrix structure. In situ demineralization of mineralized turkey leg tendon and bone. *Calcified Tissue International*, 59:474 – 479.

Rao, W. and Boehm, R. (1974). A study of sintered apatites. *Journal of Dental Research*, 53(6):1351 – 1354.

Reilly, D. and Burstein, A. (1974a). The elastic modulus for bone. *Journal of Biomechanics*, 7:271 – 275.

Reilly, D. and Burstein, A. (1974b). The mechanical properties of cortical bone. *Journal of Bone and Joint Surgery*, 56A(5):1001 – 1022.

Reilly, D. and Burstein, A. (1975). The elastic and ultimate properties of compact bone tissue. *Journal of Biomechanics*, 8:393 – 405.

Reilly, G. and Currey, J. (2000). The effect of damage and microcracking on the impact strength of bone. *Journal of Biomechanics*, 33:337 – 343.

Rho, J.-Y., Kuhn-Spearing, L., and Zioupos, P. (1998). Mechanical properties and the hierarchical structure of bone. *Medical Engineering & Physics*, 20:92 – 102.

Riggs, C., Vaughan, L., Evans, G., Lanyon, L., and Boyde, A. (1993). Mechanical implications of collagen fibre orientation in cortical bone of the equine radius. *Anatomy and Embryology*, 187:239 – 248.

Roschger, P., Gupta, H., Berzlanovich, A., Ittner, G., Dempster, D., Fratzl, P., Cosman, F., Parisien, M., Lindsay, R., Nieves, J., and Klaushofer, K. (2003). Constant mineralization density distribution in cancellous human bone. *Bone*, 32:316 – 323.

Rydberg, K. (2001). Energy efficient water hydraulic systems. In *Proceedings of the Fifth International Conference on Fluid Power Transmission and Control*, pages 440 – 446, Zhejiang, China.

Salencon, J. (2001). *Handbook of Continuum Mechanics – General Concepts. Thermoelasticity.* Springer, Berlin, Germany.

Salgado, A., Coutinho, O., and Reis, R. (2004). Bone tissue engineering: State of the art and future trends. *Macromolecular Bioscience*, 4:743 – 765.

Sanahuja, J. (2008). *Impact de la morphologie structurale sur les performances mécaniques des matériaux de construction: application au plâtre et à la pâte de ciment [Impact of the structural morphology on the mechanical performance of building materials: application to plaster and cement paste].* PhD thesis, École Nationale des Ponts et Chaussées, Marne-la-Vallée, France. In French.

Sanahuja, J. and Dormieux, L. (2005). Résistance d'un milieu poreux à phase solide hétérogène [Strength of a porous medium with a heterogeneous solid phase]. *Comptes Rendus Mecanique*, 333:818 – 823. In French.

Sanahuja, J., Dormieux, L., Meille, S., Hellmich, C., and Fritsch, A. (2008). Micromechanical explanation of elasticity and strength of gypsum: from elongated anisotropic crystals to isotropic porous polycrystals. *Journal of Engineering Mechanics*. Submitted for publication.

Sasaki, N. (1991). Orientation of mineral in bovine bone and the anisotropic mechanical properties of plexiform bone. *Journal of Biomechanics*, 24:57 – 61.

Schreiber, E., O.L., A., and Soga, N. (1973). *Elastic constants and their measurement*. McGraw-Hill, New York.

Schwefel, H. (1977). *Numerische Optimierung von Computer-Modellen mittels der Evolutionsstrategie*. Birkhäuser, Basel, Switzerland. In German.

Sedlin, E. and Hirsch, C. (1966). Factors affecting the determination of the physical properties of femoral cortical bone. *Acta Orthopaedica Scandinavica*, 37:29 – 48.

Seilacher, A. (1970). Arbeitskonzept zur Konstruktionsmorphologie [Concept for structure-morphology]. *Lethaia*, 3:393 – 396. In German.

Shareef, M., Messer, P., and van Noort, R. (1993). Fabrication, characterization and fracture study of a machinable hydroxyapatite ceramic. *Biomaterials*, 14(1):69 – 75.

Sietsema, W. (1995). Animal models of cortical porosity. *Bone*, 17(4):297S – 305S.

Silyn-Roberts, H. and Sharp, R. (1986). Crystal growth and the role of the organic network in eggshell biomineralization. *Proceedings of the Royal Society of London, Series B*, 227(1248):303 – 324.

Simo, J. and Taylor, R. (1985). Consistent tangent operators for rate- independent elastoplasticity. *Computer Methods in Applied Mechanics and Engineering*, 48:101 – 118.

Su, X., Sun, K., and Landis, W. (2003). Organization of apatite crystals in human woven bone. *Bone*, 32:150 – 162.

Suchanek, W. and Yoshimura, M. (1998). Processing and properties of hydroxyapatite-based biomaterials for use as hard tissue replacement implants. *Journal of Materials Research*, 13(1):94 – 117.

Suquet, P., editor (1997a). *Continuum micromechanics*. Springer, Wien – New York.

Suquet, P. (1997b). Effective behavior of nonlinear composites. In Suquet, P., editor, *Continuum micromechanics*, pages 197 – 264. Springer, Wien – New York.

Suvorov, A. and Dvorak, G. (2002). Rate forms of the Eshelby and Hill tensors. *Int. J. Solids Structures*, 39:5659 – 5678.

Tadic, D. and Epple, M. (2003). Mechanically stable implants of synthetic bone mineral by cold isostatic pressing. *Biomaterials*, 24:4565 – 4571.

Tai, K., Ulm, F.-J., and Ortiz, C. (2006). Nanogranular origins of the strength of bone. *Nano Letters*, 6(11):2520 – 2525.

Taylor, D. (2003). How does bone break? *Nature Materials*, 2:133 – 134.

Taylor, D., O'Reilly, P., Vallet, L., and Lee, T. (2003). The fatigue strength of compact bone in tension. *Journal of Biomechanics*, 36:1103 – 1109.

Taylor, W., Roland, E., Ploeg, H., Hertig, D., Klabunde, R., Warner, M., Hobatho, M., Rakotomanana, L., and Clift, S. (2002). Determination of orthotropic bone elastic constants using FEA and modal analysis. *Journal of Biomechanics*, 35:767 – 773.

Thelen, S., Barthelat, F., and Brinson, L. (2004). Mechanics considerations for microporous titanium as an orthopedic implant material. *Journal of Biomedical Materials Research A*, 69A:601 – 611.

Tong, W., Glimcher, M., Katz, J., Kuhn, L., and Eppell, S. (2003). Size and shape of mineralities in young bovine bone measured by atomic force microscopy. *Calcified Tissue International*, 72:592 – 598.

Traub, W., Arad, T., and Weiner, S. (1989). Three-dimensional ordered distribution of crystals in turkey tendon collagen fibrils. *Proceedings of the National Academy of Sciences of the USA*, 86:9822 – 9826.

Ulm, F.-J., Constantinides, G., and Heukamp, F. (2004). Is concrete a poromechanics material? a multiscale investigation of poroelastic properties. *Materials and structures*, 37(1):43–58.

Ulm, F.-J., Delafargue, A., and Constantinides, G. (2005). Experimental microporomechanics. In Dormieux, L. and Ulm, F.-J., editors, *CISM Vol. 480 – Applied Micromechanics of Porous Media*, pages 207 – 288. Springer, Wien.

Urist, M., De Lange, R., and Finerman, G. (1983). Bone cell differentiation and growth factors. *Science*, 220:680 – 686.

Verma, D., Katti, K., and D., K. (2006). Bioactivity in *in situ* hydroxyapatite-polycaprolactone composites. *Journal of Biomedical Materials Research A*, 78A:772 – 780.

Vesentini, S., Redaelli, A., and Montevecchi, F. (2005). Estimation of the binding force of the collagen molecule-decorin core protein complex in collagen fibril. *Journal of Biomechanics*, 38:433 – 438.

Viceconti, M., Taddei, F., Van Sint Jan, S., Leardini, A., Cristofolini, L., Stea, S., Baruffaldi, F., and Baleani, M. (2008). Multiscale modelling of the skeleton for the prediction of the risk of fracture. *Clinical Biomechanics*, 23:845 – 852.

Vitale-Brovarone, C., Verné, E., Robiglio, L., Appendino, P., Bassi, F., Martinasso, G., Muzio, G., and Canuto, R. (2007). Development of glass-ceramic scaffolds for bone tissue engineering: Characterisation, proliferation of human osteoblasts and nodule formation. *Acta Biomaterialia*, 3:199 – 208.

Walsh, J., Brace, W., and England, A. (1965). Effect of porosity on compressibility of glass. *Journal of the American Ceramic Society*, 48:605 – 608.

Weinbaum, S., Cowin, S., and Zeng, Y. (1994). A model for the excitation of osteocytes by mechanical loading-induced bone fluid shear stresses. *Journal of Biomechanics*, 27:339 – 3660.

Weiner, S., Arad, T., Sabanay, I., and Traub, W. (1997). Rotated plywood structure of primary lamellar bone in the rat: Orientation of the collagen fibril arrays. *Bone*, 20:509 – 514.

Weiner, S. and Wagner, H. (1998). The material bone: structure - mechanical function relations. *Annual Review of Materials Science*, 28:271 – 298.

Wilson, E., Awonusi, A., Morris, M., Kohn, D., Tecklenburg, M., and Beck, L. (2006). Three structural roles for water in bone observed by solid-state NMR. *Biophysical Journal*, 90:3722 – 3731.

Woodard, J., Hilldore, A., Lan, S., Park, C., Morgan, A., Eurell, J., Clark, S., Wheeler, M., Jamison, R., and Wagoner Johnson, A. (2007). The mechanical properties and osteoconductivity of hydroxyapatite bone scaffolds with multi-scale porosity. *Biomaterials*, 28(1):45 – 54.

Yosibash, Z., Trabelsi, N., and Milgrom, C. (2007). Reliable simulations of the human proximal femur by high-order finite element analysis validated by experimental observations. *Journal of Biomechanics*, 40:3688 – 3699.

Yunoki, S., Ikoma, T., Monkawa, A., Ohta, K., Kikuchi, M., Sotome, S., Shinomiya, K., and Tanaka, J. (2006). Control of pore structure and mechanical property in hydroxyapatite/collagen composite using unidirectional ice growth. *Materials Letters*, 60:999 – 1002.

Zaoui, A. (1997a). Matériaux hétérogènes et composites [Heterogeneous materials and composites]. Lecture Notes from École Polytechnique, Palaiseau, France, in French.

Zaoui, A. (1997b). Structural morphology and constitutive behavior of microheterogeneous materials. In Suquet, P., editor, *Continuum micromechanics*, pages 291 – 347. Springer, Wien – New York.

Zaoui, A. (2002). Continuum micromechanics: Survey. *Journal of Engineering Mechanics (ASCE)*, 128(8):808 – 816.

Zhong, Z. and Meguid, S. (1997). On the elastic field of a spherical inhomogeneity with an imperfectly bonded interface. *Journal of Elasticity*, 46(2):91 – 113.

Zimmermann, R. (1991). Elastic moduli of a solid containing spherical inclusions. *Mechanics of Materials*, 12:17 – 24.

Zioupos, P., Currey, J., Casinos, A., and Buffrénil, V. D. (1997). Mechanical properties of the rostrum of the whale *mesoplodon densirostris*, a remarkably dense bony tissue. *Journal of Zoology, London*, 241:725 – 737.

MIX
Papier aus verantwortungsvollen Quellen
Paper from responsible sources
FSC® C105338

Printed by Books on Demand GmbH, Norderstedt / Germany